The Logic of Life

François Jacob

THE LOGIC OF LIFE
A History of Heredity

translated by Betty E. Spillmann

VINTAGE BOOKS
A Division of Random House, New York

FIRST VINTAGE BOOKS EDITION, February 1976

English translation copyright © 1973 by Betty E. Spillmann

All rights reserved under International and Pan-American
Copyright Conventions. Published in the United States by
Random House, Inc., New York, and simultaneously in
Canada by Random House of Canada Limited, Toronto.
Originally published in the United States by Pantheon Books
in 1974. Originally published in France as *La logique du vivant:
une histoire de l'hérédité* by Editions Gallimard, Paris.
Copyright © 1970 by Editions Gallimard. English translation
first published in Great Britain as *The Logic of Living Systems:
A History of Heredity* by Allen Lane, a division of Penguin
Books Limited, London.

Library of Congress Cataloging in Publication Data

Jacob, François, 1920-
The logic of life.

Translation of La logique du vivant.
1. Genetics—History. 2. Biology—History.
3. Life (Biology) I. Title.
[QH428.J313 1976]   575.1   75-28139
ISBN 0-394-71903-4

Photo Credit: Dr. Landrum B. Shettles

Manufactured in the United States of America

Do you see this egg? With it you
can overthrow all the schools of theology,
all the churches of the earth.

DIDEROT
Conversation with d'Alembert

Contents

Contents

Introduction

The Programme

Few phenomena in the living world are so immediately evident as the begetting of like by like. A child soon comes to realize that dog is born of dog and corn comes from corn. Mankind early learnt to interpret and exploit the permanence of forms through successive generations. To cultivate plants, to breed animals, to improve them for food or to domesticate them, all require long experience. This already implies a certain notion of heredity and its uses. To obtain good crops, it is not sufficient to wait for the full moon or offer up sacrifices to the gods before sowing; it is also necessary to know how to select the right kind of seed. Farmers of the prehistoric era were somewhat like Voltaire's hero, who undertook to wipe out his enemies with a judicious mixture of prayers, incantations and arsenic. Particularly in the living world, it proved most difficult to separate the arsenic from the incantations. Even when the virtues of the scientific method had become solidly established for the study of the physical world, those who studied the living world continued to think of the origin of living beings in terms of beliefs, anecdotes and superstitions for several generations. Relatively simple experiments suffice to make short work of the notion of spontaneous generations and impossible hybridations. Nevertheless, some aspects of the ancient myths concerning the origin of man, of beasts and of the earth persisted, in one form or another, until the nineteenth century.

Heredity is described today in terms of information, messages and code. The reproduction of an organism has become that of its constituent molecules. This is not because each chemical species has the ability to produce copies of itself, but because the structure of macromolecules is determined down to the last detail by sequences of four chemical radicals contained in the genetic heritage. What are transmitted from generation to generation are the 'instructions' specifying the molecular structures: the architectural plans of the future organism. They are also the means of executing these plans and of coordinating the activities of the system. In the chromosomes received from its parents, each egg therefore contains its entire future: the stages of its development, the shape and the properties of the living being which will emerge. The organism thus becomes the realization of a programme prescribed by its heredity. The intention of a psyche has been replaced by the translation of a message. The living being does indeed represent the execution of a plan, but not one conceived in any mind. It strives towards a goal, but not one chosen by any will. The aim is to prepare an identical programme for the following genera-tion. The aim is to reproduce.

An organism is merely a transition, a stage between what was and what will be. Reproduction represents both the beginning and the end, the cause and the aim. With the application to heredity of the concept of programme, certain biological contradictions formerly summed up in a series of antitheses at last disappear: finality and mechanism, necessity and contingency, stability and variation. The concept of programme blends two notions which had always been intuitively associated with living beings: memory and design. By 'memory' is implied the traits of the parents, which heredity brings out in the child. By 'design' is implied the plan which controls the formation of an organism down to the last detail. Much controversy has surrounded these two themes. First, with respect to the inheri-tance of acquired characters. The idea that the environment can in-fluence heredity represents a natural confusion between two kinds of memory, genetic and mental. That is an old story going back at least as far as the Old Testament. To avoid further misunderstand-ings with his father-in-law, Jacob tried to develop flocks of sheep easily distinguishable by their markings. He took rods of green poplar and pilled white strakes in them and set them in the watering troughs ... that the animals should conceive when they came to drink. 'And the flocks conceived before the rods and brought forth cattle ringstraked, speckled and spotted.' Throughout the centuries, experi-ments of this sort were repeated *ad infinitum*, but not always with the same success. For modern biology, the special character of living beings resides in their ability to retain and transmit past experience. The two turning-points in evolution – first the emergence of life, later the emergence of thought and language – each corresponds to the appearance of a mechanism of memory, that of heredity and that of the mind. There are certain analogies between the two systems: both were selected for accumulating and transmitting past experience, and in both, the recorded information is maintained only as far as it is reproduced at each generation. However, the two systems differ with respect to their nature and to the logic of their performance. The flexibility of mental memory makes it particularly apt for the transmission of acquired characters. The rigidity of genetic memory prevents such transmission.

The genetic programme, indeed, is made up of a combination of essentially invariant elements. By its very structure, the message of heredity does not allow the slightest concerted intervention from without. Whether chemical or mechanical, all the phenomena which contribute to variation in organisms and populations occur without any reference to their effects; they are unconnected with the organism's need to adapt. In a mutation, there are 'causes' which modify a chemical radical, break a chromosome, invert a segment of nucleic acid. But in no case can there be correlation between the cause and the effect of the mutation. Nor is this contingency limited to mutations alone. It applies to each stage in the formation of an individual's genetic heredity, the segregation of the chromosomes, their recombination, the choice of the gametes which play a role in fertilization and even, to a large extent, to the choice of sexual partners. There is not the slightest connection between a particular fact and its consequences in any of these phenomena. Each individual programme is the result of a cascade of contingent events. The very nature of the genetic code prevents any deliberate change in programme whether through its own action or as an effect of its environment. It prohibits any influence on the message by the products of its expression. The programme does not learn from experience.

Design is another notion which has always been intuitively associated with the organism. As long as the living world appeared as a system regulated from without, as long as it was presumed to be administered externally by a supreme power, neither the origin nor the finality of living beings raised any difficulties; they remained merged with those of the universe itself. However, after the establishment of physics as a science at the beginning of the seventeenth century, the study of living beings found itself faced with a contradiction. Since that time there has been increasing opposition between the mechanistic interpretation of the organism on one hand, and the evident finality* of certain phenomena, such as the development of the egg into an adult, or animal behaviour, on the other. Claude Bernard summed up the paradox in these words:

Even if we assume that vital phenomena are linked to physico-chemical manifestations, which is true, this does not solve the question as a whole, since it is not a casual encounter between physico-chemical phenomena which creates each being according to a predetermined plan and design ... Vital phenomena certainly have strictly defined physico-chemical conditions, but at the same time they are subordinated and succeed each other in sequence according to a law laid down in advance; they are repeated over and over again in ordered, regular and constant manner, harmonizing with each other, with a view to achieving the organization and growth of the individual, animal or plant. There is a kind of pre-established design for each being and each organ, so that, considered in isolation, each phenomenon of the harmonious arrangement depends on the general forces of nature, but taken in relationship with the others, it reveals a special bond: some invisible guide seems to direct it along the path it follows, leading it to the place which it occupies.[1]

Not a word of these lines needs to be changed today: they contain nothing which modern biology cannot endorse. However, when heredity is described as a coded programme in a sequence of chemical radicals, the paradox disappears.

Everything in a living being is centred on reproduction. A bacterium, an amoeba, a fern – what destiny can they dream of other than forming two bacteria, two amoebae or several more ferns? If there are living beings on earth today, it is because other beings have reproduced with desperate eagerness for two thousand million years or more. Let us imagine an uninhabited world. We can conceive the establishment of systems possessing certain properties of life, such as the ability to react to certain stimuli, to assimilate, to breathe, or even

* Translation of the French word '*finalité*' used to describe both a finalistic or purposeful behaviour and the principle of final causes viewed as operative in the universe.

to grow – but not to reproduce. Can they be called living systems? Each represents the fruit of long and laborious elaboration. Each birth is a unique event, without a morrow. Each occasion is an eternal recommencement. Always at the mercy of some local cataclysm, such organizations can have only an ephemeral existence. Moreover, their structure is rigidly fixed at the outset, incapable of change. If, on the contrary, there emerges a system capable of reproduction, even if only badly, slowly, and at great cost, that is a living system without any doubt. It will spread wherever conditions permit. The more it spreads, the greater its protection from catastrophe. Once the long period of incubation is over, the system becomes established by the repetition of identical events. The first step is taken once and for all. In such a system, however, reproduction, which is the very cause of existence, also becomes its purpose. It is doomed to reproduce or disappear. Some beings have succeeded each other throughout a vast number of generations without changing. Some annual plants have remained unchanged for millions of years, and therefore through at least as many successive cycles. *Limulus*, the king crab of the sea-shore, is still identical with its ancestor found in the fossils of the Secondary geological era: during all this time the programme has not varied, each generation punctually fulfilling its task of exactly reproducing the programme for the following generation.

If, however, an event occurs in the system which happens to 'improve' the programme and facilitate in some way the reproduction of certain descendants, the latter naturally inherit the ability to multiply more effectively. The very finality of the programme thus transforms certain changes in programme into adaptation factors. For variability is an inherent quality in the very nature of living systems, in the structure of the programme, in the manner in which it is recopied by each generation. Modifications in the programme occur at random. It is only afterwards that a sorting operation takes place, by the very fact that every organism which appears is immediately put to the test of reproduction. The famous 'struggle for survival' merely represents a contest for progeny – an endless competition recommencing with each generation. In this eternal contest there is only one criterion – fecundity. The most prolific automatically win through the subtle interplay between populations and environment. By dint of always inclining towards those with the most offspring, reproduction ends by diverting populations along extremely precise paths. Natural selection represents only the regulation imposed on the multiplication of organisms by their environment. If the living world is evolving in the opposite direction to the inanimate world, steering not towards disorder but towards increasing order, it is thanks to this requirement imposed on living beings to reproduce, always more, always better. The necessity for reproduction, the very fact which would inevitably lead an inert system to disintegration, becomes a source of novelty and variety in the living world.

The notion of programme enables a clear distinction to be drawn between the two domains of order which biology tries to establish. Contrary to what is often imagined, biology is not a unified science. The heterogeneity of its objects, the variety of its techniques and the divergent interests of its practitioners, all lead to intellectual diversity. At the extremes are two great tendencies, two attitudes in fundamental opposition. The first may be called integrationist or evolutionist. Not only does it claim that the organism cannot be separated into its components, but also that it is often useful to consider it as an element of a system of higher order – group, species, population or ecological family. Evolutionary biology is concerned with communities, behaviour, the relationships which organisms set up with one another or with their environment. In fossils it looks for traces of present-day living forms. Impressed by the incredible diversity of beings, it analyses the structure of the living world, seeks the cause of existing characteristics and describes the mechanism of adaptations. Its aim is to define the forces and factors which have made the fauna and flora of today. For the integrationist, the organ and the function are interesting only as part of a whole that comprises not simply the organism, but the species, with its impedimenta of sexuality, prey, enemies, communication and rites. The integrationist refuses to believe that *all* the properties, behaviour and performances of a living being can be explained by its molecular structures alone. For him, biology cannot be reduced to physics and chemistry, not because he desires to invoke mystically a vital force, but because integration

6

confers on systems at all levels properties which their elements do not possess. The whole is not just the sum of its parts.

The opposing attitude may be called tomist or reductionist. For the reductionist, the organism is indeed a whole, but a whole which must be explained by the properties of its parts alone. He is interested in organs, tissues, cells and molecules. Reductionism seeks to explain functions by means of structures alone. Conscious of the unity of composition and functioning found behind the diversity of living beings, it sees the organism's performances as the expression of its chemical reactions. The reductionist believes that the components of a living being must be isolated and studied under controlled conditions. By varying these conditions, repeating the experiments, defining each parameter, this biologist attempts to master the system and eliminate the variables. His aim is to break up the complexity as far as possible and to study the components with that degree of purity and precision characteristic of experiments in physics and chemistry. For him, there is no property of the organism which cannot ultimately be described in terms of molecules and their interactions. Certainly there is no question of denying the phenomena of integration and emergence. Without doubt the whole may have properties which its components lack, but these properties always result from the structure of the components themselves and their arrangement.

The wide disparity between these two attitudes is obvious. There are differences not only in method and objective, but also in language, concepts and thus even in causal explanations of the living world. The integrationist is interested in remote causes that involve the history of the earth and of living beings for millions of generations. The reductionist, in contrast, is concerned with proximate causes that affect the components of the organism, its functions and its reactions to environment. Much of the controversy and misunderstanding, particularly with regard to the finality of living beings, is caused by a confusion between these two attitudes. Each tries to establish a system of order in the living world. For one, it is the order which links beings to one another, sets up relationships and defines speciations. For the other, it is the order between the structures by which functions are determined, activities coordinated and the organism integrated. One considers living beings as the elements of a vast system embracing the whole earth. The other considers the system formed by each living being. One seeks to establish order between organisms; the other within each organism. The two kinds of order meet at the level of heredity, which constitutes the order of biological order, so to speak. If the species are stable, it is because the programme is scrupulously re-copied, sign by sign, from one generation to another. If they vary, it is because from time to time the programme changes. On the one hand, it is necessary to analyse the structure of the programme, its logic and its execution; on the other, to examine the history of pro-grammes, their drift and the laws governing their changes throughout generations in terms of ecological systems. In all cases, however, the finality of their reproduction justifies both the structure of the present-day living systems and their history. The smallest organism, the smallest cell, the minutest molecule of protein is the result of endless experimentation unremittingly pursued through geological time. What significance could there be in a mechanism regulating the pro-duction of a metabolite by a cell, if not an economy of synthesis and energy? Or in the effect of a hormone on the behaviour of a fish, if not the protection of its offspring? There is a definite purpose in the fact that a haemoglobin molecule changes shape according to oxygen pressure; in the production of cortisone by a cell of the suprarenal gland; in the registration by a frog's eye of the forms moving in front of it; in the mouse fleeing from the cat; in the male bird parading in front of the female. In each case there is a property which confers an advantage on the organism in the competition for descendants. To adjust a response with reference to a potential enemy or to an eventual sexual partner is to adapt, in the most precise sense of the word. In evolution, a genetic programme which makes such reactions auto-matic is certain to prevail over one which does not. The same is true of a programme which allows learning and adaptation of behaviour by various regulatory systems. In each case, reproduction acts as the main operative factor: on one hand, it provides an aim for each organism; on the other it gives a direction to the aimless history of organisms. For a long time, the biologist treated teleology as he would a woman he could not do without, but did not care to be seen with in public. The concept of programme has made an honest woman of teleology.

The aim of modern biology is to interpret the properties of the organism by the structure of its constituent molecules. In this sense, modern biology belongs to the new age of mechanism. The programme is a model borrowed from electronic computers. It equates the genetic material of an egg with the magnetic tape of a computer. It evokes a series of operations to be carried out, the rigidity of their sequence and their underlying purpose. In fact, these two kinds of programme differ in many ways. First in their properties: one can change at will, the other cannot. In a computer programme, information is added or deleted according to the results obtained; the nucleic-acid structure, on the contrary, is inaccessible to acquired experience and remains unchanged throughout generations. The two programmes also differ in the role they play and in their relationships with the executive organs. The machine's instructions do not deal with physical structure or component parts. The organism, on the contrary, determines the production of its own components, that is, the organs which are to execute the programme. Even if a machine capable of self-reproduction could be built, it could only make copies of what it actually is. Since all machines wear out in time, little by little the offspring would necessarily become less perfect than the parents. In a few generations, the system would drift further and further towards statistical disorder. Descendants would be doomed to die out. The reproduction of a living being, on the contrary, does not simply involve making a copy of the parent at the time of procreation. A new object is created, starting a series of events which lead from an initial state to the state of the parents. Each generation begins from a vital minimum, that is, a cell. The programme contains all the operations which the cycle accomplishes, leading the individual from youth to death. Furthermore, the genetic programme is not rigidly laid down. Very often it only sets the limits of action by environment, or merely gives the organism the ability to react, the power to acquire some extra information which is not inborn. Phenomena such as regeneration or modifications produced in the individual by environment certainly indicate some degree of flexibility in the expression of the programme. As organisms become more complex, as their nervous system increases in importance, genetic instructions provide them with new potentialities, such as the ability to remember or to learn. However, the programme plays a role even in these phenomena – in learning, for example, it determines what may be learnt and at what stage in life learning should take place – in memory, it limits the nature, number and permanence of recollections. The rigidity of the programme thus varies according to the operations. Certain instructions are carried out literally. Others are expressed by capacities or potentialities. However, in the end the programme itself determines its degree of flexibility and the range of possible variations.

*

This book deals with the history of heredity and reproduction. It deals with the changes in the way man has looked at the nature of living beings, their structure and their continuity. For a biologist, there are two different ways of examining the history of his science. Firstly, it may be considered as a succession of ideas, thus involving a search for the thread which guided thought along the path to current theories. This is reverse history, so to speak, which moves back from the present towards the past. Step by step, the forerunner of the current hypothesis is chosen, then the forerunner of the forerunner, and so on. This view of history permits ideas to acquire independence: they behave somewhat like living organisms, being born, reproducing and dying. Endowed with an explanatory value, they can spread and invade. Such a view presupposes a sort of evolution of ideas, for which the selective forces are either a constantly improving series of hypotheses and their applications or some kind of rational finalism. Considered in this fashion, spontaneous generation, for example, was weakened by Francisco Redi's experiments. It lost more ground with those of Spallanzani and finally disappeared with those of Pasteur. However this does not explain why Spallanzani's experiments failed to carry conviction when a century later Pasteur's repetition of these experiments with only slight extensions quickly settled the issue in a conclusive fashion. The same is true of the theory of evolution. Lamarck can be seen as a predecessor of Darwin, Buffon of Lamarck,

Benoît de Maillet of Buffon, and so on. But this fails to explain why Lamarck's ideas were almost totally neglected at the beginning of the nineteenth century by men such as Goethe, Erasmus Darwin and Geoffroy Saint-Hilaire, the very people who were seeking arguments in favour of transformism.

The alternative approach to the history of biology involves the attempt to discover how objects become accessible to investigation thus permitting new fields of science to be developed. It requires analysis of the nature of these objects, and of the attitude of the investigators, their methods of observation, and the obstacles raised by their cultural background. The importance of a concept is defined operationally in terms of its role in directing observation and experience. There is no longer a more or less linear sequence of ideas, each produced from its predecessor, but instead a domain which thought strives to explore, where it seeks to establish order and attempts to construct a world of abstract relationships in harmony not only with observations and techniques, but also with current practices, values and interpretations. Ideas repudiated in time gone by often assume as much significance as those recognized by present-day science, and obstacles are as important as open pathways. Here knowledge works on two levels. Each period is characterized by a range of possibilities defined not only by current theories or beliefs, but also by the very nature of the objects accessible to investigation, the equipment available for studying them and the way of observing and discussing them. It is only within this range that reason can manoeuvre. It is within these fixed limits that ideas operate, are tested and come into conflict. Among all the possible terms, the one which best integrates the results of investigations must be singled out. This is where individual choice comes in. However, in this endless discussion between what is and what might be, in the quest for a chink revealing another possibility, the margin of freedom of the individual investigator is sometimes very narrow. The importance of the individual decreases as the number of practitioners increases: if an observation is not made here today, it will most frequently be made somewhere else tomorrow. For a long time, men have wondered what would have become of scientific thought if Newton had been an apple-gatherer, Darwin a sea-captain or Einstein a plumber (as he said he would have preferred to be). At worst, there might have been a few years' delay in the development of the theories of gravity or relativity, and even less in the development of the theory of evolution, which Wallace put forward at the same time as Darwin. When a theory appears too soon, as in the case of Mendel's laws of inheritance, no one pays attention to it. When it becomes timely for a small group of specialists, it may come to light in several places at the same time. On the contrary, once they are accepted, scientific theories contribute more than anything else to reorganizing the domain of the possible, modifying ways of looking at things, bringing to light new relationships or objects; in short, changing the existing order.

This way of considering the evolution of a science such as biology is completely different from the preceding one. There is no longer any question of finding the royal road of ideas, retracing the confident march of progress towards what now appears to be a solution, using present-day rational values to interpret the past and examine it for pointers to the present. On the contrary, it means specifying the various stages of knowledge, defining the transformations and revealing the conditions which enable objects and interpretations to enter the field of the possible. The elimination of the doctrine of spontaneous generation then no longer appears an almost linear process, leading from Redi to Pasteur by way of Spallanzani. Darwin is no longer merely Lamarck's son and Buffon's grandson. The disappearance of the doctrine of spontaneous generation and the appearance of a theory of evolution become the products of mid-nineteenth-century thought as a whole. The development of concepts of life and of the history of nature were preconditions. Before these changes could occur, there had to be a delimitation of the species, a discrimination between the organic and the inorganic and an elimination of the series of transitions leading imperceptibly from the simplest organisms to the most complex. In the end, by their rigidity and their dogmatism, their obstinacy in considering only the fixity of species, Linnaeus and Cuvier contributed at least as much to the eradication of spontaneous generation as did Redi and Spallanzani by their experiments. And by shattering the old myth of a chain of living beings, Cuvier perhaps did more to make a theory of evolution possible than did Lamarck, with his generalization of eighteenth-century transformism.

There are many generalizations in biology, but precious few theories. Among these, the theory of evolution is by far the most important, because it draws together from the most varied sources a mass of observations which would otherwise have remained isolated; it unites all the disciplines concerned with living beings; it establishes order among the extraordinary variety of organisms and closely binds them to the rest of the earth; in short, it provides a causal explanation of the living world and its heterogeneity. The theory of evolution may be summed up essentially in two propositions. First, that all organisms, past, present or future, descend from one or several rare living systems which arose spontaneously. Second, that species are derived from one another by natural selection of the best procreators. Considered as a scientific theory, the theory of evolution has the worst failing of all: founded on history, it cannot be directly verified in any way. Nevertheless, it has a scientific character, unlike magic and religion, because it remains open to contradiction by experimental evidence. Formulating this theory means running the risk of being contradicted one day by some scientific observation. However, up till now, most biological generalizations only reflect and confirm certain aspects of the theory of evolution. This is particularly true of a whole series of propositions, such as: all living beings are made up of cells; all living beings utilize the same optical isomers; the genetic information of an organism is contained in deoxyribonucleic acid (DNA); the energy a living being requires is provided by reactions in which phosphorylations are coupled with the utilization either of a chemical compound or of light. What physiology and biochemistry have shown during the present century is principally the unity of composition and functioning in the living world. Over and above the diversity of forms and the variety of performances, all organisms use the same materials for carrying out similar reactions, as if the living world as a whole always used the same ingredients and the same recipes, originality being introduced only in the cooking and the seasoning. It must be admitted that once she had found the best recipe, nature abided by it throughout evolution. Whatever his speciality, whether he deals with organisms, cells or molecules, every biologist today sooner or later has to interpret the results of his investigations in the light of the theory of evolution. Other biological theories, such as the theories of nervous conduction and of heredity, are generally extremely simple and involve only a very minor degree of abstraction. Even when an abstract entity such as the gene appears, the biologist will not rest until he has replaced it by material components, particles or molecules, as if, in order to last, a biological theory had to be based on a concrete model.

Perhaps what brought about the greatest transformation in the study of living beings is the fact that new objects have become accessible to investigation – not always as a result of a new technique for increasing sensorial powers, but often as the result of new ways of looking at the organism, examining it and formulating questions to be answered by observation. Indeed, a mere change of lighting often makes an obstacle disappear, or brings into view some aspect of an object, some hitherto invisible relationship. No new instrument was required, at the end of the eighteenth century, in order to compare a horse's leg with a man's leg, and to reach conclusions about their similarities of structure and function. Between the hand of Fernel, who coined the word physiology, and that of Harvey, who made possible experiments on the circulation of blood, the scalpel did not change in shape or quality. From all those who studied heredity during the nineteenth century up to Mendel, there was no more than a slight difference in the choice of experimental objects, in what was examined and, above all, in what was disregarded. If Mendel's work remained neglected for more than thirty years, it was because neither professional biologists, stockraisers nor horticulturists were yet in a position to adopt his attitude. 'Those who seek God, find him,' said Pascal – but they find only the God they are looking for.

Even the sudden appearance of an instrument which increases the discriminatory powers of the senses is never more than the practical application of an abstract concept. The microscope is the reutilization of physical theories of light. Moreover, it is not enough to 'see' a hitherto invisible body to turn it into an object for investigation. When Leeuwenhoek first contemplated a drop of water under a microscope, he discovered an unknown world: swarming forms, living beings, a whole unpredictable fauna which the instrument suddenly made accessible to observation. Yet thought of that period did not know what to make of this world: it had no use for these microscopic beings, no way of relating them to the rest of the living world. This discovery merely provided a topic of conversation. That they can live and move about, these beings so small that the eye cannot discern them – there was yet another demonstration (if one were needed) of the power and generosity of Nature. A subject of entertainment as well for the courts and salons given up to scientific gossip. A subject of scandal, finally, for men like Buffon, who saw a flagrant insult to the whole living world in these microscopic beings. It was an insult to all creatures, and particularly to the noblest of them all, that a drop of water could contain thousands of living beings. When, in the same period, Robert Hooke looked at a piece of cork under a microscope, he discerned honeycomb shapes which he baptized cells. Malpighi and others found similar shapes in sections of certain plant organs. However, they were unable to draw the slightest conclusion as to the structure of plants. At the end of the seventeenth century, it was still a question of investigating the visible structure of living beings, not breaking them down into sub-units. The only field in which thought was prepared to welcome the revelations of the microscope was that of generation. Hitherto, the events which accompany the mixing of seeds and the development of the egg had remained hidden through lack of adequate sensorial equipment. Thus when Leeuwenhoek and Hartsoeker observed 'animalcules' frantically swimming around in the spermatic fluid of many different types of male animals, they immediately found a use for them – but not the right one. For many decades, naturalists tried either to turn these animalcules into the sole artisans of generation, or to reduce their role to that of mere bystanders. For an object to be accessible to investigation, it is not sufficient just to perceive it. A theory prepared to accommodate it must also exist. In the dialogue between theory and experience, theory always has the first word. It determines the form of the question and thus sets limits to the answer. 'Chance favours only the prepared mind,' said Pasteur. Chance, here, means that the observation was made by accident and not in order to verify the theory. Yet the theory which enabled the accident to be interpreted already existed.

*

Like other sciences, biology today has lost many of its illusions. It is no longer seeking for truth. It is building its own truth. Reality is seen as an ever-unstable equilibrium. In the study of living beings, history displays a pendulum movement, swinging to and fro between the continuous and the discontinuous, between structure and function, between the identity of phenomena and the diversity of beings. From these oscillations, the architecture of the living gradually emerges, revealed in ever deeper layers. In the living world as elsewhere, the question is always to 'explain the complicated and visible by the simple and invisible', according to Jean Perrin's expression. However, with living beings as with inanimate objects, there are wheels within wheels. There is not one single organization of the living, but a series of organizations fitted into one another like nests of boxes or Russian dolls. Within each, another is hidden. Beyond each structure accessible to investigation, another structure of a higher order is revealed, integrating the first and giving it its properties. The second can only be reached by upsetting the first, by decomposing the organism and recomposing it according to other laws. Each level of organization thus brought to light leads to a new way of considering the formation of living beings. After the sixteenth century, a new organization is seen to have appeared on four separate occasions, each time a structure of a higher order: first, at the beginning of the seventeenth century, the arrangement of visible surfaces, which might be called the first-order structure; then at the end of the eighteenth century, the 'organization', the second-order structure which underlies organs and functions and was finally broken down into cells; next at the beginning of the twentieth century, chromosomes and genes, the third-order structure hidden in the heart of the cell; finally, in the middle of the present century, the nucleic acid molecule, the fourth-order structure, which is considered today to be the basis of each organism, of its properties and its permanence throughout generations. Each of these structures became in turn the focus of investigations.

What this book attempts to describe are the conditions which, since the sixteenth century, have enabled the successive disclosure of these structures: it is the way in which 'generation', a new creation each time and always requiring the intervention of some external force, has become 'reproduction', the intrinsic property of all living systems; it is the access to such increasingly concealed objects as cells, genes, nucleic-acid molecules. The discovery of each 'Russian doll', the demonstration of these consecutive levels are not the result of a mere accumulation of observations and experiments. More often they express a deeper change, a new way of considering objects, a trans-formation in the very nature of knowledge.

I

The Visible Structure

In 1573, in the book entitled *Des Monstres et Prodiges* which completed his treatise on generation, Ambroise Paré observed: 'Nature always tries to create its own likeness: a lamb with a pig's head was once seen because the ewe had been covered by a boar.'[1] What now surprises us in this sentence is not primarily the notion of a monster that blends the characteristics of different species; everyone has imagined or sketched one. Nor is it the way the monster was produced; once the possibility of such an exchange of forms and organs between animals has been accepted, copulation seems by far the simplest means of producing such a hybrid. The truly disconcerting feature is the argument used by Paré. To demonstrate what today appears to be one of the most regular of natural phenomena, the formation of a child in the image of his parents, he invokes the *sight* of something we consider impossible, something, indeed, which appears to us incompatible with the very regularity of this phenomenon. Unfortunately, Paré does not tell us what the descendants of the lamb with the pig's head looked like. Nowhere can we learn whether it begot other lambs with pigs' heads.

At this time, it was not even conceived that natural phenomena, the generation of animals as well as the movement of heavenly bodies, could be governed by laws. No distinction was made between the necessity of phenomena and the contingency of events. For if horse was obviously born of horse and cat of cat, this was not the effect of a mechanism that permitted living beings to produce copies of themselves, somewhat as a printing machine produces copies of a text. Only towards the end of the eighteenth century did the word and the concept of reproduction make their appearance to describe the formation of living organisms. Until that time living beings did not reproduce; they were engendered. Generation was always the result of a creation which, at some stage or other, required direct intervention by divine forces. To explain the maintenance of visible structures by filiation, the seventeenth century came to refer the formation of all individuals belonging to the same species to a series of simultaneous creations, carried out on the same model at the origin of the world. Once created, the future beings then awaited the hour of their birth, sheltered from fantasy and irregularity. However, until the seventeenth century, the formation of a being remained immediately subject to the will of the Creator. It had no roots in the past. The generation of every plant and every animal was, to some degree, a unique, isolated event, independent of any other creation, rather like the production of a work of art by man.

Generation

From ancient times to the Renaissance, knowledge of the living world scarcely changed. When Cardan, Fernel or Aldrovandus speak of organisms, they are more or less repeating what Aristotle, Hippocrates or Galen had already said. In the sixteenth century, each mundane object, each plant and each animal can always be described as a particular combination of matter and form. Matter always consists of the same four elements. An object is thus characterized by form alone. For Fernel, when an object is created, it is the form that is created.[2] When the object perishes, only the form disappears, not the matter, since if matter itself vanished, the world would have disappeared long ago; it would have been used up. The hand that confers form on matter to create stars, stones or living beings is that of Nature. However, Nature is merely an executive agent, an operative principle working under God's guidance. When one sees a church or a statue, one knows perfectly well that an architect or a sculptor exists or had existed in order to bring those objects into being. In the same way, when one sees a river, a tree or a bird, one also knows that a supreme creative Power exists and, having decided to make a world, arranges it, keeps it in order and constantly directs it.

The likeness which Ambroise Paré invokes to explain the formation of the lamb with the pig's head did not have the same status as today. In order to know things then, it was necessary to detect the visible signs which nature had placed on their surfaces precisely to permit man to comprehend their relationships. It was necessary to discern the system of resemblances, the network of analogies and similitudes providing access to certain of nature's secrets. For, said Porta, 'divine intentions may be inferred from the resemblances between things.'[3] In order to know an object, none of the analogies by which it is linked to things and to beings should be neglected. There are plants that look like hair, eyes, grasshoppers, hens, frogs or serpents. Animals are mirrored in the stars, in plants, in stones where, said Pierre Belon, 'Nature has taken more pleasure in expressing the shape of fish than that of other animals.'[4] Moreover, resemblances which are particularly difficult to discern carry a mark: they are signed. The signatures help to discover the analogies which might otherwise escape notice. Thanks to similitudes and signatures, it is possible to slip from the world of forms into that of forces. Through analogies, 'the invisible becomes visible',[5] said Paracelsus. For the resemblances are neither useless nor unwarranted. They are not the expression of mere playfulness from heaven. Certain bodies look alike because they have the same qualities. Conversely, similarity expresses common qualities. The resemblance of a plant to the eye is just the sign that it should be used for treating diseases of the eyes. The very nature of things is hidden behind the similitudes. Thus the resemblance of a child to its parents is only a special aspect of all those by which beings and things are secretly linked.

The order in a living being is no different, therefore, from that which reigns in the universe. All is nature and nature is one, as witnesses this passage which Paracelsus devoted to physicians.

The physician should know what is useful and what is harmful to unfeeling creatures, to sea monsters and fish; what is pleasing and what is hateful to animals bereft of reason; what is healthy for them and what is unhealthy. This is what he must learn about Nature. What else? The powers of magic formulas, their origin and source, their nature: who is Melusina; who the Siren is; what is permutation, transplantation and transmutation; how to grasp them and how to understand them perfectly; what surpasses nature, species, life; the nature of the visible and of the invisible, of sweet and bitter; what has a good taste; what death is; what is used by the fisherman, the leather-worker, the tanner, the dyer, the blacksmith, the wood carver; what goes into the kitchen, in the cellar, in the garden; what concerns weather, the hunter's art and the miner's trade; the life of the vagabond and the homelover, the needs of the countryside and the causes of peace; the interests of the layman and of the cleric; the occupations and the nature of different states, their origin; the nature of God and of Satan, poison and antidote, feminine nature and masculine nature; the difference between women and maidens, between yellow and buff; what is white, black, scarlet and grey; the reason for multiplicity of colours, for short and for long, for success and for failure; and how to obtain all these results.[6]

A living being could not then be reduced to the visible structure alone. It represented a link in the secret network tying together all the objects in the world. Each animal, each plant was viewed as a sort of protean body extending not only to other beings, but also to the stones, the stars and even to human activities. It had to be seen not only as it was in reality, but also in the kitchen, in the sky, on coats of arms, in the view of the apothecary, the tanner, the fisherman, the hunter. When Aldrovandus deals with the horse, he describes its shape and appearance in four pages, but he needs nearly three hundred pages to relate in detail the horse's names, its breeding, habitat, temperament, docility, memory, affection, gratitude, fidelity, generosity, ardour for victory, speed, agility, prolific power, sympathies, diseases and their treatment; after that the monstrous horses appear, the prodigious horses, fabulous horses, celebrated horses, with descriptions of the places where they won glory, the role of horses in equitation, harness, war, hunting games, farming, processions, the importance of the horse in history, mythology, literature, proverbs, painting, sculpture, medals, escutcheons.

In that period, the living world could not be arranged according to forms alone. The disposition of beings operated at a different level, according to a different cleavage of knowledge. All appeared continuous in nature, with a hierarchy rather than categories. There was, of course, Aristotle's old arrangement, the obvious difference by which living beings could be distinguished from minerals and which only the soul could account for. Among the living beings distinction could be made between plants, animals and man whose various qualities agreed with the various kinds of souls God placed in them. But in the hierarchy of beings, progression occurred by imperceptible degrees. Among these forms with overlapping characteristics, it was very difficult to decide where each domain begins and ends. A sponge – who can say it is a plant or an animal? Coral – is it really a rock? 'Just as zoophytes resemble both animals and plants,' said Cesalpinus, 'so do mushrooms belong both to plants and inanimate objects.'[7] In fact, the only segregation in the living world which the sixteenth century did not hesitate to acknowledge was between man and 'brute beasts'. The distinction between plant and animal could be made only through an area of comparison in which the evidence of differences was effaced by the importance of similarities. 'All parts of plants', said Cardan, 'correspond to parts of animals, roots are similar to the mouth, the lower parts of the trunk to the belly, leaves to hair, bark to hide and skin, wood to bones, veins to veins, nerves to nerves, the matrix to entrails.'[8] The search for resemblances blurs the portraits and fills in the gaps. The plant becomes an upturned animal, head downwards. Cesalpinus places the heart, abode of the soul, 'where we believe there is most reason for placing the vital principle . . . in the lower part of the plant, where the stem joins the root'.[9]

Amid this tangle of forms, there is no place for the species, as this is to be understood in the seventeenth and eighteenth centuries – namely, a persistence of the visible structure through succeeding generations. The production of its like by a living organism does not express a necessity of nature. To explain the production of an organism, it is necessary to resort each time to the action of God or his agents. The creation of a living being, like that of everything else, requires the union of matter and form. In addition, the properties of living beings call for direct intervention by the ruling forces of the world. The connection is ensured by two intermediaries: firstly, the soul, specific to each individual, of a quality determined by its place in the hierarchy of beings, and not perceptible to the senses; secondly, innate heat, common to all living beings, and perceptible.

The existence of a soul was as necessary then to account for the properties of living beings as electricity is now to explain thunderstorms. The most important event in generation thus became the implantation of the soul in the matter of the body. That was the 'natural', or as we would say now 'biological', event *par excellence*. As for innate heat, it was the very mark of life itself. When death occurred, heat faded away and the body grew cold, although it preserved its shape for a time. 'We recognize our friend although his life is no longer there and his heat is no longer there. The innate heat has fled,'[10] said Fernel. This heat filled all living beings, even 'the serpent, although his temperament is cold', even 'the mandrake and the poppy and all plants of cold temperament'.[11]

The Creator had placed this heat, source of all life, in two locations. One in animals and plants endowed with the ability to beget their like – more specifically, in the male seed, able to activate and mould the matter contained in the female seed. Thus, Montaigne said, 'We see women, quite alone, bringing forth shapeless lumps of flesh, who with a different kind of seed would bear good and natural offspring.'[12] The other location was in the sun, whose heat could directly activate the elements, earth, water and all kinds of debris to produce vile beings: 'serpents, grasshoppers, worms, flies, mice, bats, moles and everything that is born spontaneously, not from seed, but from putrid matter and filth',[13] said Fernel. In the eyes of the sixteenth century spontaneous generation was at least as natural and understandable as generation by seed, if not more so. Only the perfection of the forms could justify the complexity of the processes. 'Nature would have generated all animals from putrid matter,' wrote Cardan,

but as perfect beings need much time for completion, matter could not be preserved long enough without movement and principally without some conceptacle, owing to changes in weather; for these reasons, matter was necessary where the covering of the egg or fruit was kept until it became perfect, and therefore generation is accomplished by sowing of seed.[14]

To describe generation, the sixteenth century used the images, if not the models, borrowed from two of man's creative activities: alchemy and art. The use of heat to transform matter constituted the alchemists' cardinal method. When they searched for new combinations of mercury, sulphur and saltpetre, it was in the heat of furnaces and alembics. Or, when putrefaction transformed a piece of meat into a mess of flies, it was by the heat it generated. Again, when the seed of perfect animals was produced, it was thanks to the heat of the body. The matter and the spirits living in it were kneaded, ground and dispatched from the heart to the liver, from the liver to the brain, from the brain to the testicles by 'twists and resolution and meanders like tendrils on the vine',[15] Paré explained. As they progressed through the 'tortuosities and anfractuosities of the body', humours and seed took on all virtues necessary for their future work, concupiscent, ossific, carnific, nervific and veinific virtues. The play of unknown forces was hidden behind that language. Thanks to words, the mystery of nature finally yielded slightly because some of the properties they designated were housed in the words themselves. To pronounce or to write them was to gain some access to their hidden secrets, in the same way as the detection of resemblances opened the way to the knowledge of things. However, in the end, 'Parents are merely the seat of the forces uniting matter and form,' wrote Fernel. 'Above them, stands a more powerful Workman. It is He who determines the form by breathing the breath of life.'[16] The production of a living being by the forces governing the universe resembles the production of an object by man. Armed with all its powers, nature works 'like architects, masons and carpenters who, having laid the foundations of a house or the hull of a ship, erect and build the rest of the structure'.[17] Or like 'a sculptor who extracts the form out of bronze or stone'.[18] Or even like 'a painter who draws from Life'.[19] It was this likeness drawn from nature which accounted for the resemblances between parents and children by which heredity insinuated itself into the network of analogies and similitudes. Heredity represented the share of the artist, as it were, that mixture of form, constitution and temperament but not matter, which through seed reappeared from one generation to another. 'What a wonderful thing it is' Montaigne remarked

that that drop of seed from which we are produced bears in itself the impressions, not only of the bodily shape, but of the thoughts and inclinations of our fathers! Where can that drop of fluid harbour such an infinite number of forms? And how do they convey those resemblances, so heedless and irregular in their progress, that the great-grandson shall be like his great-grandfather, the nephew like his uncle?[20]

However, the portrait from nature always remained subject to the influence of various forces which could act upon the embryo during its development: 'Woman's imagination', said Paracelsus, 'resembles divine power, its external desires imprint themselves on the child.'[21] After that, everything becomes possible. All the visions, dreams and impressions can force their way into the child in its mother's belly. Every feeling of the parent can be reflected in the child, stamp its mark on him, set its signature. 'When the peahen sits on its eggs', Fernel declared, 'if it is covered with white cloths it engenders young that are white and not varicoloured. Similarly the hen engenders chicks of different colours if she sits on eggs painted different colours ... These things have been confirmed by the observation of several persons.'[22]

Thus in the production of perfect beings which beget their like, the network of resemblances becomes duplicated. On one hand, reproduction of form and temperament is ensured by heredity. On the other hand, through sensations perceived by parents or their imagination, the product of generation remains accessible to outside influences which can imprint on the child the signature of all possible analogies. The wheel comes full circle in the interplay of similitudes. Why should the child not be reflected in the universe, since the universe is already reflected in the child before its birth? In such roundabout ways, invisible forces endlessly reflect resemblances between living beings. All combinations of forms then become possible: organs, limbs, embellishments of all living beings can be rearranged with one another. From slight variations and trifling anomalies expressing a few unimportant errors to that blending in some alembic where the parts of various living beings are interchanged to produce monsters, the passage is imperceptible. There are all the intermediates from the outgrowth with the 'shape of a cherry, a plum, a horn or a fig'[23] to the 'mare bringing forth a foal with a man's face'.[24]

Sixteenth-century descriptions of the living world are filled with all kinds of monsters. Entire books are devoted to them by Aldrovandus and Ambroise Paré and in each 'History' of living organisms, of birds or of fish, fabulous and ordinary beings rub shoulders. These monsters always reflect what exists. None resembles nothing at all.

None differs entirely from what can be seen at every turn. They simply do not resemble one single being, but two, three or more simultaneously. Their parts correspond to those of different animals: 'a monster with a bear's head and a monkey's arms', 'the man with the hands and feet of an ox', 'the child with a frog's face', 'the dog with a head like a fowl', 'the lion covered with fish-scales', 'the fish with a bishop's head' and all other conceivable combinations. Monsters always bear resemblances, but they are distorted resemblances which no longer correspond to the normal action of nature. The combinations and signs which can be deciphered no longer express the order of the world, but bear witness to the errors which can slide in. 'Monsters are things which appear contrary to the course of Nature,'[25] observed Ambroise Paré. Contrary to the *course*, but not to the *forces* of nature, as nature does not make mistakes. Although errors are sometimes committed by men, animals or even plants, although physical and moral mistakes are encountered, there are never errors that might be ascribed to principles, since principles imperturbably carry out supreme intentions. 'What we call monstrosities,' explained Montaigne, 'are not so to God, who sees in the immensity of his work the infinity of shapes which he has comprehended within it.'[26] When a girl is born with two heads or a man with 'little live snakes' in the place of hair, there has been an excess of seed. If a man is born without arms or without head, on the contrary, there has been a lack of seed. If a woman brings into the world a son with a dog's head, it is not the fault of Nature, as she 'always reproduces her counterpart', but of the woman, who has indulged in sinful acts with an animal. As to misdeeds of the imagination, could they be anything else than the result of sinful thought? Each monster is the result of iniquity and bears witness to a certain disorder: an act (or even an intention) not in conformity with the order of the world. Physical or moral, each divergence from nature produces an unnatural fruit. Nature, too, has its morality.

Knowledge based on a mixture of observation, hypotheses, reasoning, antique philosophy and principles derived from scholasticism, magic and astrology, appears today to be very heterogeneous. It nevertheless composed a perfectly coherent picture. Every observation based on sensory impressions fitted a whole where each thing and each being had its place as one part of the secret network woven by the supreme power. Knowledge could not be dissociated from faith. In its nature, the formation of a living being did not differ fundamentally from the process which sets a star revolving around the earth. Generation was only one of the recipes used daily by God to maintain the world which he had created.

Deciphering Nature

During the seventeenth and eighteenth centuries – the Classical period – the creation of a living being was still considered to come about by generation, but both the role and the status of generation had changed. In less than a century, living bodies were scraped clean, so to speak. They shook off their crust of analogies, resemblances and signs, to appear in all the nakedness of their true outer shape. It was no longer possible to put in the same plane the form of a plant or animal and the notions of it expressed by travellers, historians and jurists. What was read or related no longer carried the weight of what was seen. The visible structure of living organisms then became the object of analysis and classification. And since the primary structure of organisms is on the whole faithfully repeated from one generation to the next, generation became the means by which form was maintained through time and the permanence of the species assured. The generation of a living being could no longer be considered an isolated event, unique and independent. It had become the expression of a law which demonstrated the regularity of the universe.

For with the arrival of the seventeenth century, the very nature of knowledge was transformed. Until then, knowledge had been grafted on God, the soul and the cosmos. During the Classical period, the problem was no longer to discover the secret signs of nature's intentions. It had now become the penetration of nature: the recognition of natural phenomena and of the laws by which they are linked, in so far as these are accessible to the human mind. 'Two things alone have to be considered,' wrote Descartes, 'ourselves who know and the objects themselves which are to be known.'[27] In this new relation between man and nature, the centre of action has changed: the divine will is displaced by the human mind. The question is no longer how nature was created, but how it works now. Instead of a contemplation, an exegesis or a riddle, natural science becomes a decoding. For Galileo: 'Philosophy is written in a great book which is always open before our eyes, but we cannot understand it without first applying ourselves to understanding the language and learning the characters used for writing it.'[28] For Descartes: 'If we wish to make out some writing in which the meaning is disguised by the use of a cipher, though the order here fails to present itself, we yet make up an imaginary one, for the purpose both of testing all the conjectures we may make about single letters, words or sentences, and in order to arrange them so that when we sum them up we shall be able to tell all the inferences that we can deduce from them.'[29] For Leibniz: 'The art of discovering the causes of phenomena, or true hypotheses, is like the art of deciphering, in which an ingenious conjecture often greatly shortens the road.'[30] Henceforth, science was less concerned with the divine will secretly ruling beings and things than with the grid that had to be placed on nature to decipher its order. The ultimate encoder was replaced by the decoder. What counted was not so much the code used by God for creating nature as that sought by man for understanding it. And the two were not necessarily identical. 'If for instance', said Descartes, 'anyone wishing to read a letter written in Latin characters that are not placed in their proper order, takes it into his head to read B wherever he finds A and C where he finds B, thus substituting for each letter the one following it in the alphabet, and if he in this way finds that there are certain Latin words composed of these, he will not doubt that the true meaning of the writing is contained in these words, though he may discover this by conjecture, and although it is possible that the writer did not arrange the letters in this order, but in some other, and thus concealed another meaning in it.'[31] It is the completeness of the sense, its coherence, that determines the correctness of the code revealed. In the same way, in the interrogation of nature, the power of explanation determines the value of the hypotheses or causes invoked. The method of decoding originates in the combinative system* which constitutes the esential tool of scientific research. To establish an alphabet of thought, as Leibniz wished, complexity must be reduced to the simplicity of its component units. Just as any given number must be considered as the product of prime numbers, so any logical operation must be reduced to a combination of elements. It is the combinative power of logic which shows what is possible.

Any attempt to decipher nature and reveal its order requires certainty that the grid will not change in the course of the operation. There must be an assurance of regularity in the phenomena of nature. The intervention of any hostile force must be excluded, any 'evil demon, no less cunning and deceitful than powerful', according to Descartes, 'who has used all his artifice to deceive me'.[32] Nature can be understood only as a harmony in which the behaviour of beings and objects necessarily follows the rules of a henceforth immutable game. God might well have created the world; He might well have given it the initial impetus and decided its future development. What matters today is the fact that this development cannot be modified, that nature changes nothing in the preordained plans. Otherwise, there is no science.

Until the sixteenth century, the divine will was able to realize all the possibilities that the human mind could conceive. During the Classical period the universe becomes subject to a certain regularity, certain laws or groups of laws that no one, not even God, can change, and whose logic is linked to an order of nature. Whether the laws of nature were initially imposed by some divine decree, whether the possibility of other worlds ruled by other laws can be entertained, whether the movement proceeds from the general to the particular, or vice versa, this world exists and functions. Everything is arranged, connected and harmonized, not from without as a result of some occult force inaccessible to human reason, but from within by the chain of laws themselves. Deciphering nature means restricting analysis to phenomena alone in order to discover the laws governing them. Prime causes are thrust into the background by efficient

*Translation of the French word '*combinatoire*', which means the totality of possible combination with a given set of elements.

causes. Knowledge is no longer founded on God's word, but on man's.

The analysis of objects, the successful application of the combinative method to reveal their order and measure, requires that these objects should be represented by a system of signs. The sign is no longer the stigma placed on things by the Creator to enable man to divine his intentions. It becomes an integral part of human understanding, both a product elaborated by thought for the purpose of analysis and a tool necessary for exercising memory, imagination or reflection. Among the sign systems, mathematics is evidently supreme. By means of mathematical symbols it is possible to divide up the continuum of things, to analyse them and rearrange them in various combinations. According to Galileo the great book of nature is 'written in mathematical language and the characters are triangles, circles and other geometric figures'.[33] During most of the seventeenth and eighteenth centuries, science treated only those objects which could be expressed in mathematical language, first creating a geometric universe, and later representing it in analytical form. With Newton and the abandonment of a geometric universe, calculation lost its purely mathematical sense, but enabled the results of observation and measurement to serve for the derivation of the laws of nature. It was algebraic analysis of physical phenomena which provided the universe with its law of integration. Only the mathematical formulation of physical measurements could make respectable the intervention of another mysterious force – gravitation, whose origin was unknown, but which calculation required to link together the mechanics of heaven and of earth. If physics played a decisive role in the seventeenth and eighteenth centuries, it was not only because of the way it transformed the universe and assigned new functions to observation, experiment and reasoning. It was also because physics was the only natural science that could be expressed in mathematical language. For the words of revelation, physics substituted those of logic. In place of obscurity, ambiguity and endless exegesis of the Scriptures, it established the clarity, precision and cohesion of calculation. From Galileo to Newton, physics justified the attempt of thought to establish order in the world.

First limited to mathematical objects, the search for order progressively extended to empiric domains which at first sight appeared beyond the reach of this type of analysis. Gradually the reduction of complexity to its underlying simplicity and the rules of the combinative method applied even to what could not be directly measured. The most varied objects, substances, beings and even qualities finally lent themselves to classification. To set them in order, it was sufficient, when possible, to establish the general law permitting even heterogeneous objects or propositions to be assembled, define their classes and, within the limits of this order, examine the whole field to which the law applies. This was possible whenever a suitable system of symbols could be found for representing these objects and discovering their relationships. For, as Condillac said, if a man wants to make a calculation for himself alone, he is obliged just as much to invent signs as if he wants to communicate the list of objects or the result of calculation to someone else.[34] Imagination exists only in so far as it can be expressed by a system of signs which it has itself devised.

Mechanism

The Classical period saw a progressive sharpening of the two tendencies which were taking shape in the study of living beings: physiology, the offspring of medicine, and natural history, associated with the inventory of objects in the world. However, although natural history was able to establish itself as a science because contemporary thought favoured an analysis of visible structures, physiology still remained limited for lack of both concepts and techniques. It would doubtless be advisable to distinguish a whole series of ideological tendencies in this physiology, tendencies which vary according to the practitioners, the objects studied, the aims pursued and the phenomena observed. However, we are concerned here only with those concepts which operate in the study of the living world, and there were still few to play this role. Indeed, during that entire period, the function of living beings could be understood only in so far as it reflected the function of inanimate objects.

The seventeenth century found itself in a universe whose centre of gravity had shifted: a universe in which stars and stones obeyed the laws of mechanics. Henceforth, there were only two alternative ways to assign a place to living beings and to explain their functions. Either living beings were machines in which only shapes, sizes and movements were significant, or they remained beyond the reach of mechanical laws, in which case the attempt to find unity and coherence in the world had to be abandoned. Faced with this choice, neither philosophers, nor physicists – nor even physicians – could have a moment's hesitation: all nature was a machine, just as a machine was nature. 'It is no less natural for a clock, made of the requisite number of wheels, to indicate the hours', wrote Descartes, 'than for a tree which has sprung from this or that seed, to produce a particular fruit.'[35] According to Hobbes, an animal could be considered either as a machine or as an automaton whose limbs twitch like those of a man endowed with artificial life. That was not a metaphor, a comparison or an analogy. It was an exact identification. Stars, stones and living beings, all bodies were subject to the same laws of movement. In the study of the living world, mechanism was as natural and necessary to the seventeenth and eighteenth centuries as a certain form of vitalism would become in the following century.

Until the end of the eighteenth century there was no clear boundary between beings and things. The living extended without a break into the inanimate. Everything was continuous in the world and, said Buffon, 'One can descend by imperceptible degree from the most perfect creature to the most shapeless matter, from the best organized animal to the roughest mineral.'[36] There was as yet no fundamental division between the living and the non-living. The distinction generally made between animal, vegetable and mineral was used chiefly to establish the main categories of bodies in the world. That classification could just as well have been founded on the degree of organization of bodies, their faculty of moving or their reasoning powers, as Charles Bonnet did. The distinction would then have been between 'brute or unorganized Beings, organized and inanimate Beings, organized animate Beings and finally organized, animate and reasonable Beings'.[37] There was no clear-cut division between these different groups. 'The apparent organization of foliated or layered Stones', specified Charles Bonnet, 'such as Slate or Talc, that of fibrous stones or those composed of filaments, such as Asbestos, seem to constitute transitional stages between solid brute Beings and organized solids.'[38] Organization still represented only the complexity of visible structure. Throughout the seventeenth century and most of the eighteenth century, that particular quality of organization called 'life' by the nineteenth century was unrecognized. There were not yet functions necessary to life; there were simply organs which function. The aim of physiology was to recognize their machinery and mechanisms.

In the seventeenth century, therefore, there was no particular reason to reserve a special place for living organisms and to withdraw them from the mechanics of the universe. In animals, only those functions related to the laws of movement were accessible to investigation. This was true of the bone structure of animals and their body size which, as Galileo said, cannot be increased indefinitely, 'neither in art nor in nature' without destroying the unity and hampering the normal functioning of organs. 'I believe that a small dog could carry two or three dogs of the same size on its back, but I do not believe that a horse could carry even a single horse of the same size.'[39] This was also true of the flight of birds in which, said Borelli, some relationship must necessarily exist between the weight of the body, the wingspan and the muscular power, in order for the organism to leave the ground. 'Even if he had wings, man could never fly, for want of sufficiently strong chest muscles.'[40] It was particularly true of the circulation of blood. The fibres, said Harvey, 'are in some sort the elaborate and artful arrangement of ropes in a ship'; the tricuspid valves 'are placed, like gate-keepers, at the entrance into the ventricules'; the ventricle 'forcibly expels the blood already in motion, just as the ball player can strike the ball more forcibly and further if he takes it on the rebound than if he simply threw it'.[41] It is often claimed that Harvey contributed to the establishment of mechanism in the living world by comparing the heart to a pump and the circulation to a hydraulic system. Actually this is an inversion of the order of events. In reality it was because it works like a pump that the heart was accessible to study. It was because circulation can be analysed in terms of volumes, flow and speed that Harvey could perform with blood experiments similar to those which Galileo carried out with stones. But when Harvey attempted to grapple with the problem of generation, which does not involve this type of mechanism, he could make no headway.

In the seventeenth century, the theory of animal-machines was imposed, therefore, by the very nature of knowledge. It represented an attitude inconceivable in Fernel or in Vesalius. There had perhaps been a foretaste of mechanism among the Greeks, in Aristotle or the Atomists, for example. But theirs was of a very different character. First, because the Greeks were principally concerned with analogies for teaching purposes, whereas during the seventeenth century the important issue was the unification of the forces governing the world. Second, because Aristotle assigned to the soul the origin of all movement in a living body. For Descartes, on the other hand, the properties of objects could result only from the arrangement of matter. This was true with regard to the movements of a machine whose working parts are built and arranged for the sole purpose of providing a particular movement. Obviously, this was also true with regard to the body of an animal, in which it was useless to invoke 'any other soul, vegetative or sensitive, nor any other principle of motion and life than its blood and spirits set in motion by the heat of the fire which burns continually in its heart, and which is of a nature no different from all fires in inanimate bodies'.[42] Mechanism had, therefore, to be applied to all aspects of physiology – not only to the movement of the body and the organs, but also to 'the reception of lights, sounds, smells, tastes, heat . . . the impression of their ideas on the organ of common sense and imagination, the retention or the imprint of these ideas in the memory, the internal movements of appetites and passions'.[43] Living or inanimate, all bodies in the world were thereby placed beyond the reach of any remote interaction, any dubious relationship, any attraction or repulsion through sympathy or antipathy. No place remained for the play of magic forces. Everything became possible through the action of physical forces.

It soon became obvious, however, that the resources at the disposal of mechanism were insufficient to explain the functioning of living organisms. As their complexity was revealed, the difficulty of ascribing all their properties to mere impulses acting on pulleys, levers and hooks increased. In its initial version, mechanism could not resist the growing weight of observations. Its visualization of living beings as machines composed of cogs and wheels able to transmit only a given movement inevitably led to a search for meaning and purpose outside the machine itself. A machine can be explained only from the outside. Created for a specific purpose, it serves only to perform its task. For that reason, the attempts made during the Classical period either to accentuate or limit mechanism sprang less from the attitude of contemporary science than from metaphysics. In his description of the living world, Descartes had set apart two domains: God, who having created the world and given it initial motion no longer interferes, and human thought, whose complexity surpasses that of animals or what can be achieved in automatons – language, for example: although a magpie or a parrot or even an automaton can utter words, they do not arrange them in the form of answers, by 'showing that what they are saying is the expression of thought'.[44] These are the very points which materialism and animism sought to refute.

There were two components in the animism of the Classical period. First, a need to set a high value on the living. Living beings are always imbued with magic to a certain extent. There is a sort of fetishism attached to them. They sum up all the forces of nature. In them, matter possesses miraculous properties: it is activated, influenced, transformed. With its train of images, metaphors and sympathies, the living occupies a privileged place in the world. From the outset it is placed above all other bodies. It is always given the highest rating. In comparison, inanimate objects lose their colour and relief. The passage from objects to beings, from dust to thought, expresses a hierarchy of values as much as an increase in complexity. Phenomena are not only more complex in living beings – they are also more perfect. So special a quality must correspond to a special causality. Perfection rapidly becomes transformed into an explanatory principle. The need to set value on the living in general, and on man in particular, is expressed at this time in two kinds of anthropomorphism: hierarchy is extended towards an infinite sovereign intelligence; or, on the contrary, certain human qualities are projected on all living forms. That is clearly shown, for example, by eighteenth-century interpretations of the 'admirable' regularity of honeycombs in bee-hives. Since antiquity, man has never ceased to wonder at the architecture, regularity and symmetry of the honeycomb. Towards the end of the seventeenth century, physicians and geometricians began to examine these structures more closely. They studied the bases, measured the angles and calculated the relationships. To their general astonishment, each cell was found to correspond precisely to half the structure known to crystallographers as a rhomboidal dodecahedron. It is precisely the type of symmetry which allows the best use of the space available to the cell. Each cell is in contact with twelve others, six on the same plane, three above and three below. Each cell adheres closely to its neighbours, with no space left between them. There are two possible reactions to such an efficient arrangement: either to marvel at it, or to look for an explanation in a mechanical model. Perfection can be construed in two ways. First, like Réaumur, one can attribute man's qualities to bees. The rhomboidal dodecahedron then expresses the art of the bee, its architectural qualities and even its sense of economy. 'Convinced that the bees use the pyramidal foundation which merits preference,' wrote Réaumur, 'I suspected that the reason, or one of the reasons, which made them decide in this way was to husband the wax; that among cells of the same size with a pyramidal base, the one that could be made with the greatest economy of matter or wax was that in which each rhomboid had two angles each about $110°$ and two angles each about $70°$.'[45] The bees' thrift is therefore founded on a solid knowledge of mathematics. But in the end, such an exaltation of the bee implies a denigration of man. Without ceasing to marvel, one may, like Fontenelle, make a few reservations concerning the qualities thus attributed to bees: 'The great wonder is that the exactness of the angles greatly surpasses the forces of common geometry and only belongs to new methods founded on the theory of the Infinite. But in the end, these Bees would know too much, and their exceeding glory would be their own ruin. We must accept that an infinite Intelligence makes them act blindly under his orders.'[46]

A mathematician might allow an animal to know some elementary geometry – but calculus, never!

In contrast, the other attitude to these cells is founded on analysis which ignores perfection and puts the element of wonder in its rightful place. Shape and movement alone can justify the regularity of the structures. Then, like Buffon, one can analyse the conditions under which these geometric arrangements appear. A similar hexagonal shape is often found in mineral bodies and in crystals, particularly during their formation. But it is also sometimes found in living beings, in the stomach wall of ruminants, among bodies in the course of digestion, in certain seeds and flowers. This shape always occurs when objects of similar form are subjected to more or less equal forces from opposite directions. In certain fish, for example, scales growing at the same time impede each other. They tend to occupy the space available to the best advantage, and finally adopt this hexagonal configuration. Experiments can even be made for mechanical models in which cylindrical or spherical bodies are submitted to equal pressure. 'Fill a vessel with peas, or any other cylindrical seed,' said Buffon, 'and cover it closely after pouring in as much water as the spaces between the seeds allow; then boil the water: all the cylinders become six-sided columns. The reason, which is purely mechanical, is clear: each seed, which is cylindrical, tends to occupy as much space as possible in a given area; they therefore all become necessarily hexagonal by reciprocal compression. Each bee seeks to occupy the maximum space in a given area; since a bee's body is cylindrical, it is necessary for the cells (alveoli) to be hexagonal, for the same reason of opposing forces ... People refuse to see, or they do not suspect, that this greater or lesser regularity depends uniquely on the number and shape, and not on the intelligence of these little animals; the more there are of them, the more opposing forces there are at work also, consequently the greater mechanical constraint and the greater the enforced regularity and apparent perfection in their productions.'[47] Thereafter the shape of the cell could be considered without reference to any intelligence, which in no way detracts from the beauty of the honeycombs or the poetic quality of bees.

The other component of animism during the Classical period

expressed a reaction to Cartesian mechanism and the misuse of it, particularly when its logic was carried to the extreme, as Holbach and La Mettrie wished to do. It is absurd, said Hartsoeker, to begin the study of living beings with 'the opinion that almost everything is carried out according to the laws of mechanics alone, without the aid of a soul and a mind'.[48] At that period, animism returned to an old tradition that had been revived by alchemy and medicine. However, that form of animism was less concerned with demonstrating the existence of specific phenomena of the living than combating a tendency to materialism. Initially, it was an opposition to atheism and to the acceptance of chance as one of the forces governing the world. It was the refusal to admit that causes 'produce only random consequences by their actions',[49] as Stahl put it. The perfection of beings, their properties, their generation required an unknown principle, an x beyond all understanding. There had to be a spiritual force, a psyche to carry out divine will, since no other justification of the finality of living beings could be found. This mysterious agent was given various names: first, following tradition, the Soul, then Intelligence, and even 'Plastic Nature'. At the end of the eighteenth century the agent was to change its nature to some extent and become a 'vital force'. It was then no longer a central principle, a power established in the heart of the organism and ruling its activities: it was a particular quality of the matter that composes living beings, a principle spread throughout the body, lodged in each organ, each muscle and each nerve and endowing them with specific properties. Each part of the body then possessed a 'sentiment', an 'intuitive perception', an 'instinct' as the basis of its activities. However, although vitalism was shortly to play a decisive role in the separation of beings from things and in the creation of biology, animism in the seventeenth and eighteenth centuries was not an efficient factor in promoting knowledge. Not because animists or vitalists produced fewer observations than mechanists; but because their observations were very rarely made *because of* vitalism or in order to demonstrate a vital force. Vitalism generally came into the picture *after* observation, as an aid not to investigation but to interpretation. It was not vitalism that guided Willis's scalpel in dissecting the cerebellum and determining its connections; nor Hartsoeker's eye when he perceived animalcules in male semen under the microscope. When Swammerdam discovered the metamorphosis of insects, the fact that he attributed it to the spirit and regularity of Providence mattered little. What was important in the Classical period was first to clear away the cloud of beliefs and old wives' tales that masked the outlines of objects and events. Beings and things had to be rid of the mysterious and the marvellous; they had to be brought within the limits of the visible and the analysable – in short, transformed into objects of science. This is why mechanism, despite its limitations, represented the only attitude compatible with knowledge of that time. Even the animists used analogies dear to mechanists to describe their own procedures. 'Anyone who sets out to express a reason for the phenomena of nature,' said Hartsoeker, 'is rather like a man faced with an extremely complicated machine that he can only see and examine from the outside',[50] but who has to understand how it works. All in all, animism during the Classical period represented less a scientific approach than a philosophical and ethical attitude.

Indeed, with the advent of Newton, mechanism underwent a qualitative change and, reaching into the world of substances, gave birth to chemistry. In its picture of the inanimate world, physics combined the laws of movement and the corpuscular nature of matter. Matter was no longer a homogeneous substrate which could be divided *ad infinitum*; it was seen to consist of an infinite number of isolated particles, separated from each other and not identical. Matter and movement made up the world of Descartes; Newton was to add space, that is, a void in which particles are in motion. What maintained the particles in place and bound them together to form a coherent universe was the force of attraction. This was not a component of the universe. It had no part in its construction but between all the constituent atoms of the universe it wove a network of dependencies giving cohesion to the world. It was the concept of gravitation that provided chemists with a force to replace the astral influences which, for the alchemists, had linked metals to stars and planets. When substances were blended, they did not remain inert but displaced one another. Thus between different bodies relations could be found which made them combine more or less readily. According to Geoffroy, when two substances with 'some disposition to join together' were united, if a third appeared which had more affinity for one of them, it united with it 'by making the other loose its hold'.[51] The force connecting certain corpuscles of different nature was called 'affinity'. It was no longer a magic principle, a virtue similar to those which alchemy attributed to substances. It was a property of bodies which could be measured by determining the order in which they displace one another.

Gradually, groups were delineated, families of bodies with certain common properties, such as acids or bases. Each member of a family could form combinations with each member of another. Substances could therefore be classified like plants and, for this purpose, the same methodology had to be adopted: the classification used by Lavoisier, the operations he performed, the method he applied were the same as those used by Linnaeus. Just as with plants, so with substances, the principal property had to be recognized and the substances named in accordance with that property. According to Lavoisier, in physical science the word has to elicit the idea and the idea depicts the fact for 'these are three imprints of the same seal'.[52] As chemistry is essentially an analytical science, the nomenclature of the bodies assumes particular importance: 'An analytical method is a language and a language is an analytical method.' Until that time there had been a good deal of heterogeneity in chemical language. Certain expressions, introduced by alchemists, had a somewhat enigmatic character and their meaning was clear only to the initiated. Other terms, on the contrary, had been attributed to bodies not in accordance with their properties, but by pure accident of circumstances of discovery or appearance. Chemists talked about 'oil of tartar', 'arsenic butter' or 'flowers of zinc'. For Lavoisier it was important to introduce the spirit of analysis into chemistry and this could be done only by perfecting the vocabulary. Those simple substances that could not be broken down by chemical analysis had to be named first. 'As far as I could', wrote Lavoisier, 'I designated simple substances by simple names . . . so that they express the most general and characteristic property of the substance.'[53] Compound bodies formed by uniting several simple substances had to be designated by compound names. As the number of binary combinations increased rapidly, classifications had to be introduced to avoid confusion. 'In the natural order of ideas, the name of classes and flowers refers to the property common to a great number of individuals; in contrast the name of species suggests the properties peculiar to a few individuals.'[54] Acids, for example, consist of two substances considered as simple. One confers the property of acidity common to all; the name of the class or genus must be based on this property. The other, in contrast, characterizes a certain acid: it must determine the specific name.[55] What was true for acids also applied to other bodies, to metallic calixes, to combustible materials, etc. Substances thus became accessible to order and measure. They could be classified, named, and their properties measured. Chemistry was thus transformed into a science, with its own techniques, language and concepts.

With this development of chemistry as a modified form of mechanism, a new area of physiology was open to study. Harvey had been able to analyse blood circulation in the seventeenth century because it is the only important physiological function that depends almost exclusively on the laws of movement, because the heart is a pump and the blood a liquid subject to the laws of hydraulics. In the same way, two functions analysable with the concepts and methods of chemistry became accessible to investigation in the eighteenth century: digestion and respiration. Réaumur and Spallanzani were able to begin the study of digestion, as Réaumur said, 'it is operated by the single action of a solvent and by the fermentation it causes'.[56] Stomach juice sets up a chain of chemical reactions; it acts 'on flesh and bones something in the way *aqua regia* acts on gold'. In the same way, Lavoisier was able to understand respiration because the breathing of a bird and the burning of a candle can be considered similar objects for study: they can both be analysed by means of the same concepts, by the same methods, the same measurements. The parallel to combustion led Lavoisier to link respiration with other functions, or at least with ones that could be investigated with the concepts of physics and chemistry. He first related respiration to digestion: at the end of the eighteenth century, there was no longer fire without consumption of fuel, and 'if animals did not replace with food what they lose by breathing, the lamp would soon run short of oil and the animal perish like a lamp goes out because it lacks fuel'.[57] Then he related it to circulation, because fuel must be brought for the lamp. After that, to perspiration, because of the need to keep the temperature down in the presence of a continually burning fire. In the functioning of each organ there was thus a domain which could be studied by chemical techniques – even the brain and thought processes. 'One can find out how many pounds' weight are equivalent to the efforts of a man making a speech, a musician playing an instrument. It is even possible to estimate the amount of mechanical effort in the cogitations of a philosopher, the writings of a literary man or the composition of a musician.'[58]

For Lavoisier, an animal could be studied in terms of a machine. It was no longer a machine functioning only by shape and movement, but according to extremely varied principles, since electrical phenomena had been discovered even in the muscle of a frog's leg. The best model for describing a living body was a steam engine, with its source of heat that had to be fed, a cooling system and devices for adjusting the operations of the various parts and coordinating them. 'The animal machine', said Lavoisier, 'is governed by three main regulators: respiration, which consumes oxygen and carbon and provides heating power; perspiration, which increases or decreases according to whether a great deal of heat has to be transported or not; and finally digestion, which restores to the blood what it loses in breathing and perspiration.'[59] Thus these different domains of physiology could be investigated because they had become accessible to the methods and concepts of physics and chemistry; in exchange the analogies observed and the models used contributed to the radical transformation of the way of looking at living beings at the end of the eighteenth century. Everything fitted together in the functioning of an organism, all elements were connected and all the parts articulated. Behind the forms, the demands of physiology appeared. A living body was not merely an association of elements, a juxtaposition of working organs. It was a unified set of functions, each of which corresponded to precise requirements. Not only did the organs depend on one another, but their presence and arrangement were the result of necessities imposed by the laws of nature governing matter and its transformations. What gave living beings their intrinsic properties was the interplay of relationships secretly uniting the parts so that the whole should function. It was the organization hidden behind the visible structure. Thereby the idea became possible of a nexus of qualities peculiar to living beings: what the nineteenth century was to call 'life'.

Species

Throughout the Classical period, it was primarily by their visible structure that living beings were known and investigated. The network of comparisons was substituted for the old network of similitudes. The knowledge of things was based on their relationships, identities and differences. If the thing sought and the thing given both had 'a certain nature', comparison was simple and clear. If not, a long analysis of the objects was necessary to reveal the common nature behind the complexity of the proportions. For the purpose of analysis and comparison, however, all the qualities which the senses can recognize in objects were not considered of equal value. Only what is visible enabled the universe to be understood, for although one sees a star, one does not touch it, or taste it, or hear it. It was perhaps in its attitude to the living world that the human mind had the greatest difficulty in freeing itself from habits and ideas acquired over the centuries. It had to sweep away that multitude of images which, as Tournefort put it, 'fatigue the imagination'[60] so as to reduce living beings to the limits set by the eye of the observer. Only towards the end of the seventeenth century were all the doubtful analogies rejected, all the invisible bonds and obscure similitudes, all which, said Linnaeus, 'is not clear and obvious to the lowest capacity'.[61] Only then could natural history develop, with the visible structure of living beings as its object and their classification as its aim.

To study natural history, it was necessary first to observe and describe living beings. Describing means saying what the eye discerns in an organism while rejecting everything which 'is not obvious without the aid of a magnifying-glass'.[62] It means reducing a living being to its visible aspect and translating its shape, size, colour and movement into words. The description should neglect details, but not gloss over any of the 'peculiarities' or essential elements. It should be exact and concise. For 'It is a folly,' said Linnaeus, 'to use a great many where few words are sufficient.'[63]

Natural history demands particular qualities, therefore, both in the observer and what he observes. To be a naturalist, it is first necessary to be able to renounce set images and know how to observe. But observation is not enough. One also has to see what is important – and nothing else. The naturalist cannot make shift with examining an organism as a whole. He must analyse it, study its parts, grasp its essential features. The object to be studied must meet the requirements of investigation. Obviously a plant is easier to break down into details than an animal: it is less burdened with passions and secret signs. Always moving and quivering, an animal continually changes shape. Immobile, a plant constantly displays its form and design to the observer. Beneath the animal's outer covering lurks an area of mystery; under the hair, feathers or shell, there lie the secret and confusing world of organs and the machinery of the entrails. In a plant, in contrast, nothing is concealed. All its organs are open to view, all their uses apparent. It is quite evident, remarked Tournefort, 'that it is easier to understand the structure of a machine and remember the names of its component parts if we know the uses for each part'.[64]

At first sight, the structure of an animal or even a plant is a very complex architectural object. It is difficult to compare the forms as a whole. But when the network of resemblances and differences is established not from the organism as a whole, but from its parts after analysis, complexity becomes simplicity. What is visible in a plant stands out clearly as a set of lines, surfaces and volumes. The general structure is reduced to a collection of more or less geometric figures – provided, once again, that the qualities to be observed are correctly selected, since not all visible properties offer the same guarantee of generality. Colour, for example, is subject to wide variations from one individual to another. The description, said Linnaeus, should be 'couched only in terms of art, if these are sufficient, according to the number, figure, proportion, situation'.[65] Comparison, therefore, should not be made between this plant and that, but only between the number of stamens, the shape of the calixes, the position of the anthers, the ratio of stamens to pistils. In the end, each plant can be represented as a collection of elements in a given number and proportion. Each of these can vary *ad infinitum* and each variation join with the others in infinite combinations. Botany becomes a kind of combinative system with almost unlimited possibilities.

These assemblages of elements had to be put in order and classified – a particularly delicate task, for several reasons. First, because of the diversity of the living world: the number of known varieties, which exceeded several tens of thousands at the end of the seventeenth century, was continually increasing, and the microscope had removed all limits to the living world. The second difficulty was continuity. Until the nineteenth century, not only was there no clear barrier between beings and things, but the living world formed a continuous web. Everything there was graduated and blended. Nature did not make jumps. Between quadrupeds, birds and fish she built bridges, extending its lines in such a way that everything was drawn together, connected and interwoven. 'She sends the bat flying among the birds,' said Buffon, 'while she imprisons the armadillo under the armour of a crustacean; she moulded the whale in the shape of a quadruped, and merely truncated the quadruped in the walrus and the seal who, born on the land, plunge into the billows and join the whale, as if to demonstrate the universal kinship of all generations born from the common mother.'[66] Beings can certainly be arranged in categories, but Nature herself does not recognize classes. Between two beings of different but neighbouring types the difference is minimal, 'so that it could not be smaller,' said Robinet, 'without one being exactly a replica of the other, nor greater without there being a gap'.[67] These two beings are as close together as possible. The passage from one to another allows no intermediary, no void. For 'If between any two beings' said Charles Bonnet, 'a void existed, why would there be a passage from one to another?'[68] Between the lowest and the highest degree in the series of living beings, therefore, there is an infinite number of intermediaries. Living beings as a whole form a continuous series, an uninterrupted chain 'which we see snaking across the surface of the Globe, penetrating the unfathomable depths of the Sea, leaping into the Atmosphere, plunging into celestial Space,'[69] said Bonnet. Only ignorance conceals certain links of the chain and allows naturalists to catch a mere glimpse of faulty links. The connection between individuals of all kinds appears so close that, taken as a whole, they might well 'form a single entity, a single universal being of which they would form the parts'[70], as Adanson put it.

Finally, the third difficulty in setting order in the world was that, 'There are really only individuals in nature', as Buffon said, 'and genera, orders and classes exist only in our imagination.'[71] In the extreme, therefore, to follow nature faithfully, classification of living beings should be ramified *ad infinitum*. It should comprise as many categories as individuals. However, that would make science impossible. To study botany, one has to come to terms with nature. According to Tournefort, it is necessary 'to collect, as if in bouquets, plants that resemble one another and to separate them from those that do not'.[72] That means discerning 'dividing lines' where there seems to be continuity, finding gaps which nature appears not to recognize. However, even if the universe is not truly divided, it appears to be so. That is a sufficient justification of the attempt to classify, and it is the role of the naturalist to find the most obvious gaps. 'This order which is so necessary,' wrote Fontenelle in his *Eulogy of Tournefort*, 'was not established by Nature, who preferred magnificent confusion to the convenience of Physicists, and it is incumbent on them to form, almost despite Nature, an arrangement and a System in Plants.'[73]

In order to classify plants, they must be represented by symbols – that is, named. To give a name is already to classify. The two operations are inextricably interwoven. They are two aspects of a single combinative system which must be articulated with that of visible structures, namely, the system of surfaces and volumes to whose rearrangement plants owe their diversity. The meeting point, the pivot of what can be seen, named and classified, is the *character*. According to Linnaeus: 'A plant should be mutually known from its specific name, and the name from the plant, and both from their proper character, written in the former and delineated in the latter.'[74] Linked to the details of structure, the character constitutes the 'proper mark'[75] of the plant. It represents the trace which must persist in thought after examination and description of a plant. Indeed, to describe is to say everything, to accumulate all the visible facts. Finding the character, on the contrary, means gathering from certain individuals the common properties by which they can be distinguished from others. Above all, it is selecting among the host of visible features those particular details which remain in the mind indissolubly connected with the plant and which replace its detailed image. By considering a single character of the plant, by naming it and keeping this quality alone in mind – in short, by reducing the plant to a single character – thought frees itself from the chaos of sentient images. It can then carry out its work of classification.

Classification is always a pyramid or a hierarchy made up of sets of classes at different levels, each class including one or more subclasses at the level below. Each hierarchy can function with differing degrees of complexity: simple, as in the 'synoptic disposition',[76] in which it is reduced to a set of successive dichotomous key-words; or more elaborate, as in 'systematical' classifications in which each class on one level includes more than two classes at the level below. In the eighteenth century, the living world as a whole was arranged in a hierarchy with five levels: kingdom, class, order, genus and species. The latter group was made up of varieties, the diversity of which, according to Linnaeus, came 'only from accidental changes, generally owing to the climate, soil, heat, winds, etc'.[77] In itself the distribution of living beings at five levels represented a mere convenience. It meant that an organism was not properly classified unless it could, explicitly or otherwise, be arranged in a definite group at each level.

There were two ways of constructing this type of hierarchy: it could be carried out by a sort of logical deduction, or empirically: that is by 'systems' or by 'method'. The use of systems was much older than the use of method; it descended from Aristotle via Scholasticism. To construct a system, a certain idea must already exist of the nature of the objects to be classified and of their relationship to each other. There were, therefore, as many systems as ideas, or even botanists. The number of variables made it possible to adjust classification to the empiric data to obtain the required precision. All systems tried to discern the arrangement of characters and the logical relationship which best lent themselves to classification. As Tournefort said, 'One must resort to the art of combinations: that is, the parts of the plants must be combined together so that in the end one can select those combinations forming the generic characters which give most light and that are most consistent with experience.'[78] The combinative possibilities in the use of this or that part of a plant are then analysed, and the part giving a greater number of possible combinations than those found in nature is chosen as a basis for classification. It is the arbitrary choice of criteria which determines the divisions to be made.

On the contrary, the use of method requires no *a priori* concepts. It is sufficient to compare the objects exactly and minutely to reveal their differences. There can, therefore, be only one method. It consists of choosing a reference plant, superposing other plants on it, so to speak, and carefully noting all divergences and superfluities. 'I first made a description of each plant,' said Adanson, 'putting each part of the plant in all its details into as many separate sections; and as new Species occurred, related to those already described, I described them separately, leaving out all the resemblances and only noting the differences.'[79] The mass of resemblances thus remains as a neutral background which makes the differences stand out. From the natural grouping of differences, the dividing lines come into view. More or less accentuated according to the extent of the differences, the divisions then impose the hierarchy of classes.

Systems and method, therefore, were derived from distinct or even opposing principles. However, although the procedure differed, the language was common to both and the result similar, since both techniques allowed the same five-level hierarchy to be constructed. In both cases, there was an arbitrary choice at one stage, on which the whole structure was based in the long run. Hence a difficulty common to all classifications. 'This is the most difficult point in the history of science,' said Buffon, 'knowing how to distinguish correctly between what is real in a subject and what we include arbitrarily in considering it.'[80] For the systematicians as for the methodists, for Ray, Tournefort or Linnaeus as for Magnol, Adanson or the Jussieu brothers, the choice of a new classification was justified by one single consideration: less arbitrary decisions, more natural categories. The aim of natural history was to discover the true order in nature. For that, the 'essential' had to be distinguished from the 'accidental'.

For centuries, from the time of Aristotle and throughout the Scholastic period, the unity of a living group had been based on its 'essence', made up of the sum of 'genus' and 'differentiae'. In the Classical period the meaning and role of what was called the essence of living beings became modified, but search for the essential still remained the basis of all attempts at analysis and classification. What mattered in comparing plants were the differences in essence, not the accidental features due to chance or to variables outside the laws of nature. 'Structure', said Tournefort, 'is the essential character which distinguishes plants from one another.'[81] But everyone agreed that there was a large measure of arbitrariness even in the choice of the character. According to what was expected of it, the character could be brought into play at any point in the process of comparing plants, from the 'special mark' of a particular organ to the collection of marks taken from the whole plant. For Linnaeus, the character could be 'of three sorts: the factitious, the essential and the natural'.[82] To establish the divisions between classes, it was important to use only 'the most select essential character',[83] rejecting all accidental features due to habitat, temperature, irrigation, exposure to sun or wind, in short, everything subject to variation due to conditions of culture. The essential in a plant became that peculiarity which was imposed on it by nature and free from any external action. Being the opposite of accidental, the essential was necessarily objective. It depended, not on observation, but on the distant origins of creation. The order in which the essences of living beings were arranged was that dictated by nature, not by human reason. The fact that plants had essential parts gave the naturalist a clear conscience, as it were. It allowed him to divide the classes 'without scruple'[84] as Tournefort advised, seeking only effectiveness. And what formed the essential of a plant for

Linnaeus was the element imposed by 'the continued generation of the Species'.[85]

Although the systems and method each had an inner logic, that logic had no connection with the reality of nature. It was, therefore, necessary to turn to an external element: the permanence of the visible structure in succeeding generations. The concept of species thus arose at the end of the seventeenth century from the need of naturalists to base their classifications on the reality of nature. What made the species a privileged category was that it was based, not only on similarities between individuals, but also on the succession of generations which always produce their like. 'The specific identity of the bull and the cow,' said John Ray, 'as that of man and woman, results from the fact that they are born of the same parents, often of the same mother.'[86] Order was maintained in the structure of animals and plants because that structure was faithfully transmitted from parents to descendants with their properties as a whole. The specific form of a living being was in some way passed on in its seed and 'No species is born of the seed of another, and vice versa,' observed Ray. This was true provided all sterile hybrids and 'mules' produced by certain unnatural coupling were excluded from the species. For a species to preserve its universal character, generation had to be 'continuous, perpetual and invariable,' said Buffon, 'in a word, similar to that of other animals'.[87] Under those conditions, in the living world as in the inanimate world, processes of nature were carried out in a regular fashion: they obeyed laws, of which the species was but one expression.

Consequently, in the Classical period, natural history was based on the property of living beings to beget their like and its corollary, the concept of species. In fact the species operated on two levels to make the classification of the living world possible. First, since it was not founded solely on some arbitrary division of the visible world, but involved the regularity of nature itself, the species provided a common and universally recognized foundation for all classifications of which it was the basic unit. The species was never a source of the arguments that were provoked by the genus. It justified the effort to put order in the chain of beings, to cleave up the continuous web of the living world. Only in so far as species 'existed', was the science of living beings based not on human imagination but on natural foundations. Secondly, the permanence of species through succeeding generations ensured that the living world as it now appears was indeed an accurate reflection of what was laid down at the origin. 'Of species of plants we reckon so many as there were different and constant forms of plants created in the beginning,'[88] said Linnaeus. For natural history to find its place in the body of universal knowledge, it was not enough to establish order in the living world. Classification also had to establish a bond between the contemporary living world and its origin. The concept of species testified to the permanence of living forms since creation. 'An individual is nothing in the universe,' said Buffon,

a hundred or thousand individuals are still nothing. Species are the only beings in Nature; perpetual beings, as ancient and as permanent as Nature herself; each may be considered as a whole, independent of the world, a whole that was counted as one in the works of creation and that, consequently, is but one unit in Nature.[89]

For the taxonomic relations which can be established between organisms in terms of their visible structure are always based on an underlying set of logical premises: the permanence of forms for Tournefort and Linnaeus; the functional integration of organisms for Cuvier; or, ultimately, evolutionary filiations for Darwin.

Preformation

With the concept of species, generation became an expression of the regularity of nature. During the Classical period, however, generation could be envisaged only through the visible structure of living beings and the laws of mechanics. Unlike the circulation of the blood, the phenomena of generation could neither be investigated by measuring volume, movement and speed, nor described in terms of levers, pulleys and pumps. When Harvey devoted himself to the study of the generation of living beings, he could do no better than his teacher, Fabricius of Acquapendente, who had incubated hens' eggs and opened one every day to study the state of the embryo.

Harvey sacrificed Charles I's hinds during their mating season, opening one each day to observe the contents of the uterus. But all he could observe was a shapeless mass, a sticky little heap, a sort of 'scar' where the heart, blood vessels, intestines, a head and feet gradually became visible. Harvey was reduced to making somewhat unconvincing analogies. The female is fertilized by the male 'in the same way as iron touched by the magnet acquires magnetic power'.[90] Or the uterus looks like the brain because 'there is excited by coitus within the uterus, a something identical with . . . an imagination or a desire in the brain'.[91] Because, as an epigraph of his *Treatise on Animal Generation*, he wrote the celebrated '*Omnia ex ovo*', Harvey is often credited with the concept that every living being comes from an egg. But for Harvey, '*ovum*' denoted more than 'egg'. It was any substance already organized to some degree – putrefying meat, rotten plants, excrement, the pupa or the chrysalis of insects; in short, anything from which a living being could be seen to emerge, whether quadruped, fly, worm or plant.

In the sixteenth century, spontaneous generation had been just as easy to explain as generation by seed, since it required the direct action of divine forces on matter. In the seventeenth century, occult forces had to be replaced by the organization of matter and the laws of movement, called upon to account for the formation of living beings as well as for falling bodies or the movement of the stars. For Descartes, indeed, there was no particular difficulty in that. Since matter was identical in all the bodies in the world, living beings differed from objects only in the way that matter was arranged. A mere nothing should therefore suffice to animate matter and bring forth a living body – a little heat or pressure, slight friction to produce the reaction of the parts on each other. 'Since so little is required to make a being, it is certainly not surprising that so many animals, worms and insects form spontaneously before our eyes in all putrefying substances.'[92] Heat and movement have to act gradually on each part in turn. Neither in meat nor in the eggs of perfect animals can the little being be formed at one stroke, to spring fully armed like Minerva from the head of Jupiter. Matter is set in order progressively, organ by organ, with the regularity of a highly complex clockwork mechanism. In beings which engender their like, all the mechanics are already prepared by the arrangement of matter in the seed. When speaking of the development of the embryo, Descartes used similar terms to those later employed by Laplace to describe the movement of the universe: 'If we really knew all the parts of the seed of any particular animal, man, for example, we could deduce from that alone, by certain and mathematical reasoning, the shape and conformation of each of his limbs.'[93]

Spontaneous generation gradually succumbed to the increasing weight of observation. Progressively, as the eye, aided by a magnifying glass or a microscope, obtained a closer view of insects, the complexity of their visible structure increased. A whole network of precisely interwoven fibres, vessels and nerves was gradually revealed. 'A fly possesses as many organic parts as a horse or an ox, or perhaps even more,' marvelled Malebranche. 'A horse has only four legs and a fly has six . . . There is only one crystalline lens in an ox's eye . . . but today we find many thousands in the eye of a fly.'[94] Observations by Swammerdam and Malpighi revealed the metamorphosis of the silk-worm, the cricket, the scarab beetle and the butterfly. These two naturalists described the sexual organs and method of copulation. All these accumulated observations were incompatible with the formation of worms or flies by the heat of fermentation in meat. But it became possible to discount spontaneous generation of insects in the seventeenth century because the necessary experimentation involved only movement – that of the air and of living beings. Simply putting meat in a hermetically sealed vessel was sufficient to prevent it putrefying or producing flies. In his book on generation Francesco Redi explained that reading Homer had led him to make this kind of experiment. If the putrefaction of flesh is sufficient to engender insects, asked Redi, why in Book XIX of the *Iliad* was Achilles so anxious that the body of Patrocles should not become the prey of flies? Why did he ask Thetis to protect the body against insects which might produce worms and corrupt the flesh of the dead?[95] Experiments showed that Achilles' fears were justified. 'These worms are all engendered by insemination,' said Redi, 'and the putrefying matter in which they are found is no more than the place or nest where animals lay eggs at the time of generation and where they find their food; in other words, I assert that nothing is ever engendered from this matter.'[96] The same applies to the worms living in the intestines of man and animals. They are not born of the entrails of their hosts, but from a seed which is found in the air, in water or in food and is inhaled or swallowed by the animals. It also applies to insects born in plants. It is not the fruit, roots or galls of the plants that engender them, but insects which lay their eggs there. By the end of the seventeenth century, it was well established that worms, flies and eels were born of worms, flies or eels. Where a living being appeared, a similar being had existed to engender it. Logically, the doctrine of spontaneous generation should have then disappeared; however, it quickly took refuge in the invisible and slightly grotesque world of 'animalcules' which had suddenly been perceived in ditch-water, plant infusions and saliva by means of the microscope. Further experiments were not sufficient to dispose of it. The concept of species had first to be consolidated, its limits defined and its permanence established.

The secret of generation by parents was hidden in the seed. That was where the principles of form and movement had to be substituted for occult forces. Errors of the formative ability, whims of maternal imagination, the influence of food or dreams on the formation of the child, in short, all the fantasies and irregularities which disturb the normal course of things, were inconsistent with the harmony of the universe and the laws of nature. But the only aspect of generation which the seventeenth and eighteenth centuries could approach by analytical methods, the only one accessible to their means of observation and the resolving power introduced by the microscope, was the content of the seeds. Only there could the Classical period substitute figures and particles for principles and virtues. At that time, one of the simplest questions that could be asked was: What does the seed of each sex contain? More exactly, why do certain animals lay eggs while others produce living beings? By dint of examining, dissecting and peering into bodies, the 'testicles' of viviparous females were finally found to contain small masses filled with a liquid similar to white of egg and turning yellow after copulation. Regnier de Graaf even managed to establish a correlation between the number of these lumps and the number of embryos appearing in the uterus. Until the nineteenth century, when embryologists were to show that the lumps detected by de Graaf are in fact follicles surrounding the real eggs, they were considered to be eggs themselves. At the end of the seventeenth century, therefore, all females possessed eggs. 'I have no more doubt that the testicles of a woman are ovaries,'[97] said Steno. Whether the little animal came from an egg, like the chick, or emerged alive from its mother's belly, like the calf, the process was the same. In goats, ewes, cows, in all animals dissected, the same anatomy was always found. 'Generation occurs in the same way in women,' said de Graaf, 'since they have eggs in the testicles and tubes attached to the uterus like beasts.'[98] Despite the protestations of the blue-stockings, who were indignant at being taken for hens, endless discussion went on as to whether eggs could form without copulation, whether they existed in virgins and frigid women, and whether a woman expelled an egg each month during menstruation or, on the contrary, whether the egg broke away only under 'the lash of pleasure'.

In male semen, the microscope revealed numberless creatures just like those in ditch-water or an infusion of hay – little worm-like things living, swarming and swimming in all directions. In that hitherto unsuspected world, suddenly revealed by the microscope, only the animalcules in male semen could find a place and a purpose. It was difficult to know what to make of animalcules in ditch-water, where to place them and whether to consider them a subject of wonder or of scandal. The animalcules in male semen, on the contrary, represented almost what reason had been looking for. If the male was to play his part, if seed were to be divested of principles and virtues, then particles and organized bodies had to be found. Actually, the windfall exceeded expectation. 'There are not as many men living on the surface of the globe as animalcules in the sperm of a single male,'[99] said Leeuwenhoek, who was careful to add that the examinations had not been made at the expense of his own posterity.

So there are eggs in females and animalcules in males: and that must suffice to produce the complexity of an animal. If one rejects the notion of a formative faculty and denies the intervention of any mysterious forces, if one wishes to organize these particles and transform them into animals by the laws of movement alone, then the problem is insoluble. 'It is impossible for the union of the two sexes to produce a work as admirable as the body of an animal,' said Malebranche. 'One can well believe that the general laws of the communication of movement are sufficient to form and develop the parts of organized bodies, but it is impossible to believe that they ever form such a complex machine.'[100] At a time when living beings are known by their visible structure alone, what has to be explained about generation is the maintenance of this primary structure through succeeding generations. The structure cannot itself disappear; it has to persist in the seed from one generation to another. To maintain the continuity of shape, the 'germ' of the little being to come has to be contained in the seed; it has to be 'preformed'. The germ already represents the visible structure of the future child, similar to that of the parents. It is the plan of the future living body, not potentially, in some active part of the seed from which the body of the little being is organized progressively, in the same way as a plan is carried out; but already materialized, like a miniature of the organism to come. It is like a scale model with all the parts, pieces and details already in position. The complete, although inert, body of the future being lies already waiting in the germ. Fertilization only activates it and starts it growing. Only then can the germ develop, expand in all directions and acquire its final size, like those Japanese paper flowers which, placed in water, unwind, unfold and assume their final shape. It is the same with animals as with plants: in the seeds of many plants, indeed, the miniature form of the future plant can be distinctly discerned in detail, with the outline of the stems and branches and folded leaves. There is no need to organize matter during fertilization and embryonic development. Each living being begins as something already resembling it. In animals as in plants, the levers, pulleys and springs are sufficient to ensure the development of the germ, which is only a matter of growth. Each piece of the germ gradually expands in all directions, to remain when fully grown what it had previously been in miniature.

The main question about generation then became: Which of the two seeds, male or female, contains the germ? Obviously there can be two schools of thought. One places the germ in the egg; fertility then lies in the female, in whose ovary the miniature form of the future living being remains inert, awaiting the moment of fertilization. The male plays only a very modest part in that case, merely activating the germ in some fashion by means of the spermatic fluid. There were, in fact, a host of arguments favouring the attribution of the prime role in generation to the egg. First, there were experiments with hens' eggs, in which Malpighi and others discerned the forms of the future chick, without any incubation. The cock has no organ for intromission; he sprays semen on the eggs to activate them. The same is true of fish, where the male, again, only sprays semen on the eggs laid by the female. It also applies to frogs, whose mating antics are most peculiar: the male settles himself behind the female, clasping her in his arms; they then remain entwined for weeks on end and no one knows how the male plays his role. Swammerdam maintained that frogs are like fish and that the male sprays his semen over the eggs the female releases. However, it appeared very doubtful if fertilization could take place outside the body of the female, and in any case, it was difficult to see how the germ could be situated anywhere except in the egg. As to man, everyone knew that the liquid he releases with so much pleasure does not penetrate the uterus of the woman, but runs out of it as soon as it is discharged. Plenty of cases were known of girls becoming pregnant without even letting the man's fluid penetrate them. It was only that highly 'spirituous' part of the fluid which entered the uterus, rose into the ovaries and penetrated the egg by one of the pores which existed for the purpose. Or perhaps that 'subtle' part of the male fluid insinuated itself into the blood vessels of the woman, where it mingled with her blood, setting off 'ravages which torment women', and reached the ovary where the egg was fertilized only when all the woman's blood had been fertilized. To be sure the resemblance of children to the father might give rise to some confusion; but it was quite obvious that the force of the male, concentrated in the spirit of the seminal fluid, played a role in the organization of the foetus. The shape of the embryo depended, indeed, on various factors: the form of the germ, obviously; the mother's activity and way of life during the period of gestation; but also the force with which the parts of the embryo were activated when they were impregnated with the father's seminal spirit.

However, it was also possible to support the opposite thesis which places the germ in the animalcules that swim in the male semen. Then all fertility is attributed to the male, 'which is more in keeping with his dignity'. Why invent creatures necessary for generation in the egg, when they could be seen swimming around in male semen? The generation of living beings becomes merely the development of those animalcules, each of which, said Hartsoeker, 'contains a little male or female animal of the same species hidden under a tender and delicate skin'.[101] The role of the female is then limited to providing a nest and the nourishment necessary for the development of the animalcules. These penetrate the uterus, rise into the ovary and look around for a suitable place to take up their abode. Only one succeeds in penetrating the egg, which 'has only one opening to allow the entry of one worm', said Hartsoeker, '. . . and as soon as a single one has entered, the opening closes and does not allow another worm to enter'.[102] Once installed in its egg, the little being grows and develops imperceptibly until it reaches the size and maturity compatible with birth. By dint of closely examining the animalcules, two kinds were finally distinguished. 'Animalcules differ in sex,' according to Leeuwenhoek, 'and Male and Female can be distinguished.'[103] Many tried to discern through the microscope the shape of each animal in its animalcule; but in vain. Despite all their efforts, the shape hidden under the skin of the animalcule was not discernible. The stunted foetus could only be imagined, in the shape of a homunculus, for example, its legs drawn up and its head between the arms, sitting inside its little worm.[104] It had to be accepted. The germ looks like a fish or a grub, in keeping with nature's practices. Who would recognize the cockchafer in the grub from which it is born? Who would believe that splendid butterflies with luminous wings have begun as hideous caterpillars creeping slowly along? There is no doubt, Geoffroy concluded, 'man begins by being a Worm.'[105]

Lacking any recourse to a hidden secondary structure, the preformation of germs represented the only means of ensuring the permanence of visible structures by filiation. Nevertheless, this only moved the problem one step backward. For the true generation then became the formation of the germ in the seed. The origin and organization of the germ now had to be explained. Moreover, the difficulty remained, for to avoid referring to some mysterious power or formative virtue meant returning to the laws of movement, which are just as inadequate for organizing a germ as for organizing an embryo. Only one solution was left: to consider that the germs of all organisms past, present or future, had always existed, that they had been formed at the time of Creation and were only awaiting the moment of activation by fertilization. This was the theory of the 'pre-existence' of germs. Since all the germs had to be infinitely small, there was no question of seeing them, even under a microscope. But, according to Malebranche, 'The mind should not stop at what the eye sees, for the vision of the mind is far more penetrating than the vision of the eye.'[106]

Pre-existence can be viewed in two ways with respect to the location of germs awaiting activation. First the germs can be placed outside existing beings and spread throughout nature, as Claude Perrault suggested. The germs are too small to catch sight of, but they exist in the water, in the food men eat or the air they breathe. Then 'Generation cannot fail to take place,' said Perrault, 'because as the little bodies are almost countless in kind and species throughout the world, it is difficult for them not to occur in the homogenous substance of the seed, or not to be carried to it.'[107] Germs choose members of the same species in which to settle and form the little being which awaits fertilization in order to grow.

But germs can just as well be located inside living beings. The little preformed creature then contains the germs of its own future children, who contain the germs of their own children, and so on. In a single apple-pip, there are, according to Malebranche, 'apple-trees, apples and apple-seed for infinite or almost infinite centuries to come, in the proportion of one perfect apple-tree to an apple-tree in the seed'.[108] The same applies to animals. The germs of all possible men of all time date from the Creation, but according to Swammerdam's expression they can be contained in the loins of Eve or of Adam.[109] All depends on the place assigned to germs, either in the egg or in the animalcule. If the germ lives in the egg, then 'the females of the first animals were created with all those of the same species which they were to engender in the course of time'.[110] If, on the contrary, the germ is placed in the animalcule, then 'the first males were created with all those of the same species which they have engendered or are to engender to the end of time'.[111] In any case, both versions of pre-formationism suppose that male and female germs are different, since only one contains the germs of all its descendants, nested into one another like Russian dolls. From one generation to the next, the size of the dolls decreases in the ratio of an organism to its egg. The foetus that will be born a thousand years hence is thus as well formed as the one that will be born in nine months. Only the size is different. But the diminutiveness of the germ which makes it invisible to the human eye does not exempt it from the laws of nature. There would certainly be cause for alarm if one did not know that matter is in-finitely divisible. In principle, preformation should protect the development of the foetus from the fantasies of maternal imagination. Yet 'only a year ago,' Malebranche wrote,

a woman gazed too long at the picture of St Pius, whose Canonization was being celebrated, and she gave birth to a child who perfectly resembled the picture of the Saint. He had the face of an old man, as far as possible for a child without a beard. His arms were crossed on his breast and his eyes turned up to Heaven ... This was seen by all Paris as well as by me, as it was preserved for quite a long time in alcohol.[112]

Reduced to the knowledge of the visible structure of living beings and to the laws of mechanics, the seventeenth-century scientist was led to relegate real generation, which organizes living beings from matter, to the domain of prime causes; and these he refused to con-sider. Science was interested only in the universe as it existed for that period – that is, in the products of creation and the laws expressing the regularity of their movements. Although the generation of a living being was not yet seen as a re-production, a re-formation of the child in the image of its parents, nevertheless it had acquired a new role and status. It was no longer an isolated creation independent of others, the direct realization of an intention unconnected with similar intentions. The production of a living being became one stage in the gradual realization of a single long-term project in the course of time. For like always to produce like and for species to be maintained, the germs of individuals forming the species must all have been created together on the same pattern and at the same time. Since it was not possible then to picture each birth as the reconstruction of the visible structure under the influence of some other structure of a higher order, the only solution left was the production of successive generations by a simultaneous creation. What was destined to become the father, the son and the grandson had to be created together. The idea of pre-existence fitted in naturally with the concept of species. If the germ preformed in the seed of the begetter had been formed at the time of creation together with all those of its species, no place remained for any outside intervention during generation, for irregularities due to the fantasies of parents or sins against nature. Generations could succeed generations, always identical because they were always the result of the activation of identical products drawn from the same basic stock. The species became a collection of germs, a reserve fund of copies made on the same pattern.

Preformation and pre-existence, accordingly, placed the generation of living organisms on the same level as other phenomena of nature. Living beings, like inanimate objects, 'cannot begin except by creation,' said Leibniz, 'or end except by annihilation'.[113] The universe has sprung from God's hands, fully formed and completely equipped with all its parts. Everything has been drawn from nothing by His will. Each star, each stone, each living being to be born in the course of time has been formed by a definitive and finished creation. After that initial impetus, the system has functioned regularly according to the laws of nature and without any further divine intervention. Stars revolve, stones fall, living beings are born.

Throughout the eighteenth century and as long as living beings were looked on as combinations of visible elements, preformation and pre-existence offered the only possible solution to the problem of generation. They provided the only rejoinder to the argument of Fontenelle:

Do you say that Beasts are Machines just as Watches are? Put a Dog Machine and a Bitch Machine side by side, and eventually a third little Machine will be the result, whereas two Watches will lie side by side all their lives without ever producing a third Watch.[114]

The production of a living being thus remained the result of a project whose conception and realization could not be separated from the creation of the world. It was the visible order of living beings that was maintained by descent. The continuity of living forms in species required the continuity of these forms through the very processes of generation. How could a hen be produced from an egg, unless the egg contained the characteristics of a hen, that is, a certain visible structure? As to activation and development of the germ, it would be 'no more reasonable to ask us how this takes place,' said Haller, 'than to ask why resorption of the seed of the Male makes his beard sprout'.[115]

During the eighteenth century, however, many observations showed the meticulous attention, patience and trustworthiness of the naturalists; many experiments revealed their ingenuity and skill. Experimentation still depended on mechanistic principles and techniques. It was therefore limited to those aspects of generation accessible to its methods: for example, amputation of a limb or even cutting of a body into pieces, followed by observation of the results. The component of generation most accessible to mechanistic analysis is obviously the seed itself. The spermatic fluid can be prevented from reaching its natural destination, or collected separately and mixed with the eggs afterwards. The liquid can be diluted or heated, to find out if it still remains active. It was through such investigations that generation became accessible to experimental study, even if the techniques were still fairly primitive. However, so great remained the necessity for preformation and so inconceivable the possibility of another solution that any result obtained was interpreted as supporting the theory of a preformed germ.

The dispute over the attribution of the prime role in generation to the egg or the animalcule was soon decided in favour of the egg, as a result of the discovery of parthenogenesis. On the advice of Réaumur, Charles Bonnet began to study the multiplication of aphids. And to everyone's astonishment it was found that a single individual can produce descendants. 'Take a little one at birth,' advised Bonnet,

shut it up at once in complete solitude, and the better to ensure its virginity, take extreme precautions, become even more vigilant than Argus in the Fable: when the solitary little creature has grown somewhat, it will begin to give birth, and after a few days you will find it in the middle of a large Family.[116]

Since an organism exists which can multiply 'without having sexual commerce with any individual of its species',[117] it is difficult to see how the germ can be anywhere but in the egg. Haller and Charles Bonnet both considered that the membrane lining the inside of the yolk continues that lining the intestines of the chick embryo. However, 'Since the yolk exists in eggs which have not been fertilized, it necessarily follows that the germ exists before fertilization.'[118] Obviously, there is but one conclusion: the chick does not owe its existence to the fluid provided by the cock; its 'miniature' already existed in the egg before any fertilization took place!

As for the role of sperm animalcules in generation, it remained very uncertain. Buffon considered that these worms have a kind of animal life – but not a true animal life – and there were endless arguments as to the 'animality' of the animalcules. Even if they have animality, animalcules in male semen are not distinguishable from those born and living in ditch-water. They are to be found everywhere: not only in spermatic fluid, but also in females (for example bitches in heat), and even in raw meat, veal jelly, excrement. Far from being the fertilizing principle of the male, the animalcules might well serve to homogenize the spermatic fluid or to encourage the attraction of the seeds. Perhaps they could even serve only for 'venereal pleasure'.

The role of the spermatic fluid and the animalcules it contains can be studied in simple mechanistic terms, those of matter and force. By physical means one can prevent a transfer of matter, but not a transfer of force. Oviparous organisms were considered the most suitable for these experiments, since one can obtain both fertilized and unfertilized eggs. In particular, frogs were used because of their mating habits, in which the male, clasping the female from behind, remains intertwined with her for days on end. Réaumur's faithful collaborator was entrusted with the task of observing one of these couples and not taking her eyes off it throughout the whole performance. 'Just as I had asked her to do,' said Réaumur, 'she fixed her gaze on the backside of the male and never looked away. Almost immediately she saw a spurt of liquid which she could only compare to a puff of pipe smoke.'[119] Nothing proved, however, that the puff of smoke played the slightest role in the fertilization of eggs, clusters of which, during this time, had been liberated by the female. If there was really a transfer of matter from the male to the eggs, and if that jet corresponded to an emission of semen, it must be possible to prevent the semen from reaching the eggs, which would then remain sterile. Réaumur thought up the idea of putting 'breeches made of a bladder, well-fitting breeches sealing the backside'[120] on a male frog before mating. But fertilization was not disturbed, because the frogs managed to claw the breeches off with their legs and little tadpoles still emerged from the eggs. Spallanzani, with greater skill, succeeded in making breeches which stayed on. 'The males dressed in this way mated, but the results of the copulation were merely what could be expected – not one of the eggs hatched, because none of them had been moistened with spermatic fluid, of which I observed clearly visible little drops inside the breeches.'[121]

With this material, it became possible to obtain separately each kind of seed outside the animal's body. Hence the idea of attempting fertilization by mixing the two kinds of seed in a phial in order to 'give life to this species of animal artificially', wrote Spallanzani, 'by imitating Nature's way of multiplying Amphibians'.[122] Some days after being sprinkled with male milt, the eggs hatched, tadpoles emerged and started swimming about. In this fashion, artificial insemination could be realized with very different animals – toads, salamanders, silk-worms; even a bitch in heat, when spermatic fluid taken from a male dog was injected into the uterus with a syringe.

Artificial insemination provided the tool necessary for an experimental analysis: an observable and measurable activity. It became possible to find out whether one of the seeds, subjected to various preliminary treatments, could still play its role in fertilization. Spallanzani was thus in a position to determine what is active in

semen: some force in the spermatic fluid or the animalcules themselves. Diluted in a litre of water, a drop of frog semen was still active –
hardly surprising, in view of the enormous number of animalcules.
When placed over spermatic fluid without touching it, eggs were not
fertilized: the liquid therefore, acts by contact, and the idea of
'spermatic vapour' can be ruled out. When the animalcules were
killed, by heat for example, the seed still proved capable of fertilizing
eggs: the active principle of male seed does not therefore, lie in the
animalcules, but in 'a force which stimulates the little heart of the
Tadpoles'.[123] Only Spallanzani's last experiment was defective, and
the result was not to be corrected for more than half a century.
Spallanzani considered that it confirmed the presence of a preformed
germ in the egg, a new proof 'that the little machines of the foetus
originally belonged to females and that the male provides the liquid
which determines their movement'.[124]

Even the most direct evidence was rejected. The development of
an embryo in an egg could, indeed, be studied with the technical
means available in the eighteenth century. What could be observed
there concerned shape and movement only, since the question was to
distinguish between the development of a preformed germ and the
progressive elaboration of the young animal by 'epigenesis'. By
observing the development of the chicken under the microscope,
Caspar Frederic Wolff discerned superposed membranes, first simple,
then folded, forming the swellings, grooves and tubes from which
emerged the rough outlines of organs: the nervous system, then the
vessels, the digestive tube, etc. The primary structure of a living
being was not, therefore, preformed in the egg. It was gradually
organized by a series of pleats, swellings and blisters, a sequence of
mechanical operations in time and space. That is precisely the conclusion von Baer drew from similar observations half a century later.
Although in the nineteenth century Wolff's book *Theoria generationis*
was to provide the basis of experimental embryology, it remained
almost completely ignored during the eighteenth century. There was
no framework in which to insert epigenesis, no solution to the
question of the generation of living beings, apart from that of
preformation.

Heredity

Although they found it impossible to abandon the idea of pre-existence and preformation, eighteenth-century scientists were nevertheless able to demonstrate the inadequacy of such theories. A simple calculation was enough. According to Buffon, a single germ is more than a thousand million times smaller than a man: if man's size is taken as the basic unit, the germ's size would be expressed by the fraction $\frac{1}{1,000,000,000}$ that is, 'by a number of ten figures', that of the second-generation germ by a 'number of nineteen figures', and at the sixth generation, 'by a number of fifty-five figures'. However, compared with the size of the sphere of the universe 'from the sun to Saturn', supposing the sun to be a million times bigger than the earth and distant from Saturn by a thousand times the solar diameter, the size of the smallest atom which could be seen through a microscope would be expressed by a number of fifty-four figures. The germ of the sixth generation, therefore, would be smaller than the smallest possible atom! Preposterous, concluded Buffon.[125]

Most important also appeared the phenomena of regeneration, the ability of certain animals to produce a whole body from parts – like the aquatic worm studied by Bonnet, for example, or the hydra observed by Trembley. Like a tree putting forth branches, 'this fresh-water Polyp with arms like Horns' grows young polyps which enlarge, break away from the parent and in their turn produce young polyps. A polyp can be cut up, reduced to fragments, chopped up, so to speak: in two or three weeks, each of the pieces produces a perfect new polyp from which little polyps sprout. In the same way, once cut, the 'legs' of the crayfish grow again, like the salamander's limbs and even the head of a snail. What is more, whether the leg of the crayfish is cut off near the body or at the end, whether the whole limb is removed or only a fragment, exactly the missing amount always grows again. How can all this be reconciled with preformation? Would the amputated organism be obliged to turn to one of its potential descendants and borrow the missing part? According to Réaumur, it would have to be supposed 'that there is no place on the leg of a crayfish without an egg containing another leg; and what is more wonderful still, a part of the leg similar to the part where the egg is to the tip of the limb.'[126] Better still, once the leg has grown again, it can be cut off a second time: another one still forms. The new leg, like the first, would have to contain an infinite number of eggs for replacing exactly the parts of the leg that might be removed!

There were, finally, the phenomena of heredity. Until that time, family likenesses had not greatly worried the partisans of preformation. If a child derived features from both parents, that could always be explained by some activating or nutritive force. What began to attract attention was the regularity of likeness. When a black man marries a white woman, the two colours always blend: the child is always born with an olive skin. The child always inherits some features from the father and others from the mother – height, facial features and a number of physical and moral characteristics. The same is true of animals: when copulation of an ass and a mare is fruitful, it produces neither ass nor horse, but always a blend of the two. How could pre-existence and preformation be reconciled with the unforeseen circumstances of copulation? As Maupertuis asked: 'Would the little ass, already completely formed in the mare's egg, have asses' ears because an ass put the parts of the egg in motion?'[127] Ridiculous, concluded Maupertuis.

However, although the phenomena of heredity began to assume new importance in the eighteenth century, they were still objects of observation, not experiment, at least in animals. As long as living beings appeared as combinations of visible elements, only rearrangement of those elements could be expected from cross-breeding between animals with widely different characteristics. Even if the concept of species had eliminated all possibility of a child with a dog's head or a ewe with a fish-tail, the living world was still continuous. It still formed a flawless web in which all was linked through imperceptible transitions. The limit nature sets on sexual intercourse between animals remained undefined. Yet, an attempt could be made to verify the rumours that were still rife concerning the products of illicit mating of animals belonging to nearly allied species. For example, the formation of strange animals reputed to be the result of copulation between bull and mare, cow and ass, or bull and she-ass. Or again, the union of a dog and a she-cat, or a hen and a drake. Charles Bonnet suggested that Spallanzani should put a 'voluptuous Spaniel in the company of a Doe Rabbit'.[128] Réaumur installed a hen in a cupboard with a buck rabbit, 'who used the hen as he would have done a doe rabbit, and the hen let him do everything she would have permitted a cock to do'.[129] Despite the caresses of the 'ardent buck rabbit' and the 'strong inclination these two ill-matched animals showed for each other', the eggs the hen laid remained sterile, to Réaumur's great regret. However, all these unsuccessful experiments reinforced the idea that filiation is indeed the basis on which the notion of species must be established.

Faced with these failures, Réaumur limited his ambitions to mating animals of the same species with marked differences in certain easily recognizable features. He procured two kinds of hens: some which 'differ from all others because they have an extra part, a very long claw', others in which 'a very considerable and remarkable part is missing, the parson's nose'.[130] With that material Réaumur envisaged an experiment in which the various types would be mated according to various 'combinations'. A five-clawed hen, he explained, could be mated with a four-clawed cock, and vice versa; and similarly, animals with and without a parson's nose could be mated. If chicks were born of these unions, the very fact that they had or had not a parson's nose 'should show us whether the germ originally belonged to the female or the male'. The originality of this project lay in the idea of analysing, by means of cross-breeding, the behaviour of one or two specific traits, rather than a large number. It was precisely this approach which enabled Mendel to found a science of heredity more than a century later. For although Réaumur conceived the project of such hybridizations, he never referred to them again, nor mentioned their realization, or results. It was in plants that, once their sexuality had been revealed, a few hybrids could be produced. These turned out to share the characteristics of both parents. They have an intermediate nature, said Koelreuter, 'just as in the union of an acid salt and an alkaline salt, a third, namely a neutral salt, is formed'.[131] Some of these hybrids were fertile. In their progeny certain characteristics of the parents were seen to appear through successive generations. All that was hardly compatible with preformation.

What the eighteenth century could not yet achieve through experimentation in animals, it obtained through observation in man. Although all the fantasies about the products of coupling of man and other species had not yet been banished, peculiarities of form could be observed with greater ease and certainty in man than in animals. Observation was then no longer limited to mere listing of the features which, in a child, recall the father, or the mother, or both. It became a genealogical enquiry in which certain anatomical peculiarities were followed as far as possible through succeeding generations. 'Jacob Ruhe, surgeon in Berlin,' said Maupertuis,

was born with six fingers on each hand and foot; he had this peculiarity from his mother Elisabeth Ruhe, who had it from her mother Elisabeth Horstmann of Rostock. Elisabeth Ruhe transmitted it to four of the eight children she bore to Jean-Christian Ruhe, who had nothing extraordinary about his feet or hands. Jacob Ruhe, one of the six-digital children, married Sophie-Louise of Thüngen at Dantzig in 1733, a woman who had no unusual features; they had six children; two boys were six-digital. One of them, Jacob Ernest, had six toes on the left foot and five on the right; on the right hand he had a sixth finger, which was cut off; on the left, he had only a wart in the place of the sixth finger. From this genealogy which I have followed with exactitude, six-digitism is seen to be transmitted by both the father and the mother.[132]

The same anomaly was observed in a Maltese family described by Réaumur[133] and Charles Bonnet.[134] There again the results of the enquiry led to the same conclusions. The language and methods of mathematics could even be applied to the phenomena of heredity. To exclude the effect of chance on repetition of this anatomical abnormality within one family, Maupertuis relied on the calculation of probabilities. In a town of 100,000 inhabitants, enquiry revealed the presence of six-digitism in two people only.

Let us suppose—although it is unlikely—that three others escaped me and that out of 20,000 men, one six-digital man is to be found; the probability of his son or daughter not being six-digital is 20,000 to 1; and that his son and grandson will not be six-digital is 20,000 times 20,000 or 400,000,000 to 1: finally, the probability that this peculiarity will not continue for three consecutive generations is 8,000,000,000,000 to 1; a number which is so great that the certainty of best-demonstrated things in physics does not approach this degree of probability.[135]

In talking about living beings, the physicist used the language of mathematics and the reasoning which he applied to inanimate objects. If the laws of chance were valid in one case, they could not be ignored in the other.

*

During the eighteenth century, the study of living organisms passed little by little from physicians to professionals of a new kind, naturalists. However, it had not yet acquired its individuality nor found its own methods, concepts or even language. The results of taxonomy had installed order in the chaos of visible forms; and the progress made by physiology began to reveal a hidden order within the depths of living beings. But visible order and hidden order still belonged to different domains; there was as yet no point of contact between them. Natural history in the eighteenth century constructed a fresco, a two-dimensional outline, a chart into which the living world could be inserted. Not until the end of that century, and particularly during the following one, would the organism acquire a new dimension and fresh depth. Then new relationships were to be established between the surface of a living being and its inside, between the organ and the function, the visible and the invisible.

With the concept of generation, neither the formation of a living organism nor the persistence of species could escape final causes. Even after the change of thinking which withdrew the motion of the world from control by occult forces and entrusted it to the laws of nature, even when these laws became responsible for the development of the germ, it was still necessary to resort to individual creation for each living being, past, present or future. But to invoke the pre-existence of germs was, in some way, to recognize the impossibility of accounting for generation by the sole effect of the laws of movement acting on passive matter. Through the Classical period, the methods of experimental science used by physics hardly applied to a living world which still remained permeated by anecdotes, beliefs and superstitions. Although the concept of the species brought together like with

71

like in the course of time, its boundaries still remained ill-defined. Many monsters had disappeared, but not all. In spite of Réaumur's efforts, there were still those 'abominable passions' described by Voltaire which led to rearrangements of organs, fabulous beings and fantastic genealogies. As for hybridizations, the limit between the possible and the impossible still remained hazy. At the same time, however, the example of physics and the successes in unifying celestial and terrestrial mechanics enhanced the weight of observation and experiment in deciphering nature. There was a growing tendency to adopt the attitude upheld by Newton in the introduction to his *Opticks*. 'My design in this book is not to explain the properties of Light by Hypotheses, but to propose and prove them by Reason and Experiments.'[136] The logic of *a priori* systems was replaced by the logic of facts. And in the study of the living world, certain facts, such as the regeneration of worms and polyps or the likeness of a child to its father *and* mother, did not fit well with the existence of a preformed being in the germ.

The concept of *reproduction* was born of all these observations. The expression was first used to designate the phenomena of regeneration of amputated limbs in certain animals. The part restored is the part which existed before amputation. If a crayfish's leg is cut off, the leg is re-generated, re-formed, re-produced. The word first appears, it seems, in a memoir by Réaumur, published in 1712 among the works of the *Académie des Sciences*[137] and entitled *Concerning Various Reproductions Occurring in Crayfish, Lobsters, Crabs, etc. And among others, that of their legs and shells*. That sense was to be preserved throughout the whole of the eighteenth century, particularly in the works of Charles Bonnet. In the article 'Reproduction' in the *Grande Encyclopédie*, we find: 'By reproduction is ordinarily meant the reproduction of a thing which existed previously and which has been destroyed since. Example: the reproduction of the limbs of the crayfish.' It was apparently Buffon who gave a wider meaning to the expression. In his *Natural History of Animals*, published in 1748, the term 'reproduction' was applied, not only to the re-formation of amputated parts, but to the generation of animals. Chapter II, entitled 'Concerning Reproduction in General', begins: 'Let us examine more closely this property common to animals and plants, this ability to produce its like, this chain of successive existences of individuals which constitutes the true existence of the species.'[138] Thus linked to the concept of species, the term 'reproduction' was used by everyone. Although the article 'Reproduction' in the *Grande Encyclopédie* still gave the meaning of re-forming a missing claw, the article 'Generation' indicated: 'This term is generally taken to mean the ability to reproduce, which is attributed to organized beings.' Even convinced partisans of preformation spoke of reproduction. Charles Bonnet, for example, called one chapter of his *Philosophical Palingenesis*: 'Another Feature of the Excellence of Organized Machines. Their reproductions of different genera.'[139] Although the term 'reproduction' was generally accepted, it did not have the same meaning as that understood by Buffon. For Haller, Charles Bonnet, Spallanzani and most scientists at the end of the century and even the beginning of the nineteenth, living beings were still born from pre-formed germs.

Buffon, like Maupertuis, was looking for an alternative solution to preformation to explain at one and the same time the formation of like by like, the re-generation of missing parts and the phenomena of bilateral heredity. Beyond the apparent diversity in the means of generation the question was, for Maupertuis, to discover 'the general proceedings of nature in its production and preservation'.[140] For Buffon, the important thing was to discern, behind the peculiarities, 'the mechanism which nature uses to carry out reproduction'.[141] And that mechanism could only be a hidden one. There was no way of accounting for reproduction by the persistence of visible structure. For Maupertuis, as well as for Buffon, a secret structure of a higher order was required to bind together and organize the visible parts. With the help only of Newtonian mechanics, and for lack of adequate techniques and concepts (which were not to appear until the following century), these attempts were doomed to failure. The idea of reproduction, however, the search for a mechanism common to all living beings, the necessity of penetrating the visible surface and resorting to hidden organization – all these contributed to make a new science possible: that is, biology, the science of life.

2

Organization

As long as living organisms were perceived as combinations of visible structures, preformation provided the simplest explanation for the persistence of those structures through succeeding generations. The linear continuity of the living world in space and time required a continuity of form through the actual process of generation. The role of generation was to perpetuate visible order. The species represented a sort of rigid entity, a prescribed frame into which the individual being was fitted. Filiation, therefore, had to have the same inertia as the whole system.

In the second half of the eighteenth century and the beginning of the nineteenth, the very nature of empirical knowledge was gradually transformed. Analysis and comparison tended to operate, not only on the component parts of objects, but also on the relationships between these components. It was within living bodies themselves that the very cause of their existence had to be found. It was the interaction of the parts that gave meaning to the whole. Living bodies then became three-dimensional entities in which the structures were arranged in depth, according to an order prescribed by the working of the total organism. The surface properties of a living being were controlled by the inside, what is visible by what is hidden. Form, attributes and behaviour all became expressions of organization. By its organization the living could be distinguished from the non-living. Through organization organs and functions joined together. Organization assembled the parts of the organism into a whole, enabled it to cope with the demands of life and imposed forms throughout the living world. Organization became, as it were, a structure of a higher order to which all perceptible properties of organisms were referred. Thus with the start of the nineteenth century, a new science was to appear. Its aim was no longer to classify organisms, but to study the processes of life; its object of investigation was no longer visible structure, but organization.

Memory and Heredity

As early as the mid eighteenth century, living creatures were generally referred to as 'organized beings' or 'organized bodies'. However, 'organization' still implied no more than an unusually complex arrangement of the component parts of the visible structure: the existence of a hidden structure had to be postulated only in the context of the Newtonian physical universe. In the Newtonian mechanics of matter, secret combinations of particles underlay visible combinations of surfaces and volumes. The intrinsic qualities of bodies and the properties of substances were determined, not simply by the nature of the atoms that composed them, but also by the relations of attraction or affinity between these atoms. The attributes of living organisms were, therefore, necessarily determined by the nature and interrelationships of their constituent particles. For organized beings, just as for inanimate objects, the visible structure rested on an arrangement of particles, united through the action of a force comparable to gravitational attraction, which gave coherence to the whole.

In the second half of the eighteenth century, the notion that living beings were composed of elementary units was expressed with increasing frequency. For the physiologist Haller,[1] who attempted to analyse the texture and function of muscles and nerves, an organism 'is composed partly of fibrils and partly of an infinite number of little platelets which, by their different orientations, cut off little spaces, form little areas and unite all the parts of the body'. The elementary unit of which organized bodies are built is the 'fibre'. The fibre is to the physiologist what line is to the geometrician. The 'smallest or simplest fibre, such as reason, rather than the senses, helps us to perceive',[2] represents, so to speak, the theoretical limit of anatomical analysis, resolvable by separating muscles, nerves or tendons with the point of a scalpel. There exists only one kind of fibre for all the organs. The same fibres are interwoven in all directions in a continuous web linking the parts of the body together. What gives the firmness or suppleness, elasticity or rigidity to an organ is the way in which the fibres are intertwined, the close-knit or loose arrangement of the mesh, the amount of liquid it contains. Fibres already have a complex structure and their union gives the organism its properties. 'The smallest fibre,' stated Charles Bonnet,[3]

even the tiniest fibril can be imagined as infinitely minute Machines with functions of their own. The whole Machine, the great Machine, therefore, is the result of grouping a prodigious number of 'machinules' whose actions are concurrent or converge towards a common aim.

More often than not, however, the notion of organisms being composed of elementary units represented a requirement, not of anatomy, but of logic. A crystal of sea-salt, Buffon explained, is a cube made up of other cubes and doubtless the ultimate constituents of salt are also cubes. In the same way,

animals and plants that can multiply and reproduce themselves in all their parts are organized bodies composed of other similar organic bodies, whose primary and constituent parts are also organic and similar and which we see with the naked eye in the mass, but whose primary parts we perceive only by reasoning and analogy.[4]

The reduction of organisms to a collection of units was, therefore, derived directly from the corpuscular theory of matter, and in a sense completed that theory. The elementary units composing organized beings were called 'living particles' by Maupertuis and 'organic molecules' by Buffon. According to both scientists, the units play in living beings the same role as atoms in inanimate objects. Just as the arrangement of atoms fixes both the form and the qualities of things, so the arrangement of living particles determines both the shape and the properties of beings. Like atoms, these units that the eye cannot see but which logic has to accept, represent the ultimate aim of all investigation. Like atoms, living units are linked by the force termed 'attraction' by physicists and 'affinity' by chemists, the force that gives beings and things their cohesion. Like atoms, units are indestructible. Yet the units are not atoms. They are particles of a special type, restricted to living beings alone. Ultimately, the composition of living creatures is distinguished from that of inanimate objects only by the nature of their elementary components. When an organism dies, the particles that constitute it do not perish. They simply dissociate and become available to form a new combination, to make a new being. These living particles exist everywhere in nature. They are swallowed or inhaled by organisms, which sort out the organic molecules and reject the 'brute molecules'. Living particles are first used for the development of the organism; once it has become adult, the surplus serves to form seed for its reproduction. 'The seminal fluid of each species', said Maupertuis, 'contains an innumerable multitude of parts able to form, by their assembly, animals of the same species.'[5]

For both Maupertuis and Buffon, the reproduction of living beings and their elementary composition went hand in hand. To explain generation other than by preformation, it was necessary to invoke a hidden order behind the visible order and to consider an organism no longer as an indissociable whole, but rather as a collection of 'primary and incorruptible parts' that could be joined or separated. According to Buffon, 'The assembly of these parts appears to our eyes as organized beings and consequently reproduction or generation is only a change of form that comes about and is wrought by the simple addition of these similar parts, just as the destruction of an organized being takes place through the division of the same parts.'[6] The production of a living organism costs nothing in materials: nature simply rearranges the units made available by the death of other organisms. The parental image reproduced in the generation of organized beings is an arrangement of organic molecules, a pattern of units peculiar to the species: in a word, organization.

The eighteenth century was thus able to find in the corpuscular structure of matter a solution to the *impasse* of preformation. However, it still had neither the concepts nor the technical means to investigate the hidden structure it postulated. For example, the molecules that constituted the parents and those of the seed destined to produce the child were necessarily identical. Each seed contained a complete range of the different types of particles required to form the various organs. All the parts of the body, therefore, had to participate in the production of the seed, each contributing its characteristic molecules. 'Experiment might perhaps shed light on this point,' said Maupertuis, 'if one attempted to mutilate some animals from generation to generation for a long period of time: perhaps the amputated parts would be seen to diminish gradually; perhaps in the end they would disappear.'[7] That experiment has already been carried out, retorted the partisan of preformation, Charles Bonnet: in certain tribes where it is the custom to amputate one testicle from each male, boys continue to be born with two.[8]

It was the force of attraction that assembled the particles of seed to form the child. Each part of the child was formed by the union of particles of the same type which, coming from the father and the mother, recognized each other because of the very great affinity between molecules of the same kind. Some have tried to see in this attraction between like particles from the two parents an anticipation of the specific pairing which twentieth-century geneticists observe between homologous chromosomes. However, the eighteenth and twentieth centuries are speaking of different things. For geneticists, the combination of hereditary features and their rearrangement during successive generations are mediated by independent factors that govern the expression of characters, while remaining completely distinct from them both in nature and in role. For Maupertuis, on the other hand, the particles in the seed are indistinguishable from those which are present in the body of the organism and which determine its properties. Each seed consists mainly of particles identical with those in the body of its progenitor. Naturally, therefore, the child resembles its parents: it is composed of identical particles. Since the same particles recur in the seed at each generation, characters are perpetuated by filiation. But 'Chance, or a shortage of family traits, will sometimes produce other assemblages.'[9] Such chance combinations are responsible for bizarre offspring: for instance the production of monsters, or the birth of a white child to a black couple. But if the eighteenth century could envisage a secret order governing the form and properties of a living being, it was unable to imagine different levels of structure. No distinction was made between the 'family feature', the material particle contained in the body, which determines this special feature, and the particle of seed which governs its reproduction.

Nevertheless the eighteenth century clearly perceived one of the requirements imposed by any system through which, at each generation, the form of the parents is reproduced in the child by the combination of basic particles: the intervention of a 'memory' to guide the arrangement of the particles. This problem did not arise, as long as the formation of all germs was ascribed to a simultaneous creation: for the preformationist, the 'memory' necessary to maintain visible structure through succeeding generations was in fact that structure itself. But from the moment when particles must reassociate at each generation to produce the parental image, some means of conserving this image through successive generations had to be postulated. Two possible solutions were then proposed. For Maupertuis, a follower of Leibniz, the memory directing the living particles to form the embryo is the same thing as mental memory. Matter is endowed not only with memory, but also with 'intellect, desire or aversion'.[10] The living particles are drawn together by affinity, but only their memory can explain the place they occupy in the embryo. Each particle 'preserves the memory of its former position and will return to that place whenever it can, so as to form the same part in the foetus'.[11]

For the materialist Buffon, on the contrary, what preserves the image of the parents in generation and determines the position of organic molecules in the child is not some quality common to all grains of matter, but a special structure. The particles cannot assume the parents' shape without a pattern to guide them or a mould to fashion them. 'In the same way as we can make moulds to shape the exterior of objects as we like, let us suppose that Nature can make moulds by which she shapes not only the outside but also the inside.'[12] For what is to be reproduced in forming a living being is not only that collection of lines, surfaces and volumes making up the visible shape, but also the internal arrangement, the hidden structure of organs that governs the functioning of the living body. Hence the reproduction of living beings requires what Buffon called an '*interior mould*',[13] the only way of 'imitating the interior of bodies'. Buffon has often been derided for this remark, both by his contemporaries and by his successors. In fact, he clearly perceived and analysed one of the principal obstacles to explaining reproduction and growth, an obstacle only recently overcome by molecular biology. It is possible to copy structures in one or two dimensions, but not in three.

Buffon used the mould as a model because the most obvious way of reproducing a three-dimensional body is that of the sculptor who uses a wax or plaster cast of the body. But the wax only 'perceives' the surfaces of the object, as it were. It cannot 'touch' anything below the surface and is unaware of what goes on inside the object. So the mould reproduces only the surface of the object, below which no particular order is imposed on matter during reproduction. Hence the sculptor's mould is not adequate for the reproduction of organisms: there must be an *interior* mould as well. In Buffon's words,

It may be objected that this expression, interior mould, at first sight appears to contain two contradictory ideas, that the idea of a mould can refer only to the surface and the idea of the interior mould should be related to the mass; it is as if one wanted to combine the idea of surface and the idea of mass, and one might just as well say 'massive surface' as 'interior mould'.[14]

The same difficulty is found in considering the development of the embryo, since growth of the organs takes place, not in two, but in three dimensions. Contrary to general belief, said Buffon, development cannot be achieved by mere addition of molecules to surfaces, but 'by an intussusception that penetrates the mass'.[15] It is indeed necessary for the material used in growth to penetrate into each part and into all dimensions according to 'a certain order and a certain measure, so that no more substance arrives at one interior part than another'.

The interior mould accordingly constitutes a hidden structure, a 'memory' that organizes matter so as to produce a child in the image of its parents. The child is thus formed by epigenesis. This neo-formation is not, however, an entirely new production, as Aristotle and the sixteenth century understood it – a complete organization of the being out of the chaos of matter – since the memory of the organization that existed in the parents is preserved by the continuity of the interior mould. 'What is most constant and unalterable in nature,' said Buffon, 'is the imprint or mould of each species, both in animals and plants; what is most variable and most corruptible, is the substance of which it is composed.'[16] Maupertuis, for his part, imagined the existence of a privileged class of living particles that preserve the typical characteristics of the species through successive generations.

Is this instinct [of particles], like the spirit of a Republic, spread throughout all the parts that form the body? Or, like a Monarchy, does it belong only to some indivisible part? In this case, would the indivisible part not constitute the very essence of the animal, while the others are only wrappings or sorts of clothing?[17]

In the eighteenth century, the idea of living beings composed of elements was still beyond the reach of observation and experimentation. With the means at his disposal, Buffon exercised all his ingenuity to prove the existence of organic molecules, their universal distribution throughout nature and their power of combining with each other. But his results convinced no one. Moreover, an origin had to be found for these living units as distinct from atoms. Although Buffon attributed their formation to the activation of matter by heat, final causes seemed to be almost the only explanation of their creation. For the supporters of preformation, living units represented only one more theory. Its most immediate effect was to give fresh vigour to the idea of spontaneous generation. For if organic molecules existed everywhere and could assemble under the influence of heat, was it not organic matter itself – which, Buffon said, could be 'regarded as a universal seed'[18] – that produced this invisible and extravagant world perceived under the microscope, all those little beings that did not even warrant the name of animals, swimming in rainwater, infusions of herbs and semen? There experiment could take a hand, for once again it involved a simple mechanical principle. Since heat kills animalcules, it is sufficient to seal meat juice in a vial, to heat it and see whether animalcules can still multiply or not. Abbé Needham noticed that heat did not prevent the multiplication of the tiny grubs. More meticulous, Abbé Spallanzani found no living animalcules after heating meat juice. But the effect of heat was always a debatable question, depending on whether it was considered to act on matter or on a force. 'Matter' was the presence of animalcules in meat juice or in the air of the vial: Spallanzani's experiment then ruled out the possibility of spontaneous generation, even among microscopic organisms. 'Force' was a property of meat juice or air, a 'generative force', a 'fecundity', an 'elasticity' necessary for the multiplication of the animalcules; in this case, the experiment demonstrated nothing at all. How can the existence of that force be proved? Or how can it be disproved? Not until the following century was the human intellect prepared to renounce the possibility of spontaneous generation.

In proposing their theories, neither Maupertuis nor Buffon had the least intention of dabbling in metaphysics. As good Newtonians, they wanted to base the properties of living beings on the prevailing laws of physics. Elementary units, living particles and organic molecules were invoked only to reconcile the mechanistic interpretation of the living world with the Newtonian interpretation of the universe. For Buffon, living organisms, like inanimate objects, are simply combinations of units, by which 'nature can vary her works *ad infinitum*'.[19] This is as true for living bodies as for chemical substances: only the elements possess individuality; compounds are transitory. The study of organisms then becomes a search for the laws governing the combinations of these units. Reproduction, as required by the dual heredity observed in hybrids, is merely the mechanism that makes it possible for the units to assemble. From that, a new attitude emerged during the second half of the eighteenth century. Behind the diversity of structures, processes and habits among organisms, a unity of composition and function had to be found in the entire living world. Considering animals that breathe, Buffon sees 'always the same basic organization, the same senses, the same viscera, the same bones, the same flesh, the same movement of fluids, the same play, the same action in the solids'.[20] What counted in the investigation of living bodies was not simply the organs accessible to observation: it was also the way they are linked. It was their *organization*.

The Hidden Architecture

Throughout the eighteenth century, organization still described only the combination of structures and the mosaic of elements that characterized a living being. At the end of the century, however, organization acquired a different role and status. By progressively replacing visible structure, organization provided a hidden foundation for the bare data of description, for the being as a whole and for its functioning.

On the one hand, the work of Lavoisier reversed the relative importance attributed to organs and to their functions, imposed the concept of basic functions that satisfy the needs of the whole organism and revealed the necessity of coordinating them. If respiration is always a combustion, each living being, whatever its form and habitat, must be able to obtain oxygen. It must obtain fuel from food, carry it to the place of combustion, reject the waste matter and control its temperature – in short, combine accurately a whole series of operations. The lungs, the stomach, the heart or the kidneys can no longer be considered independently. A living being is no longer a simple association of organs, each working autonomously. It becomes a whole whose parts are interdependent, each performing a particular function for the common good.

On the other hand, the attitude of naturalists was progressively changing. Throughout the eighteenth century, the anatomy of each species had been studied in isolation. The literature of that period is rich in monographs which describe in detail the structure and attributes of the lion, the bee or the bat. At the end of the century, however, anatomists no longer limited their work to describing independently each organ of a living being. They attempted to relate organ to performance and to compare the same organ in different animals or the different kinds of organs in the same animal. Investigating the leg of a horse was no longer sufficient. With Daubenton, it must be compared with a human leg, in order to define the analogies in the number of bones, their shape and their functions. Or with Camper, the brain and auditory system of fish must be brought into relation with those of man in order to link structure with function. Or again, with Vicq d'Azyr, a parallel must be drawn for different species of carnivorous animals between the structure of the teeth, the stomach, the claws and the muscles in order to reveal their relationships and show their constancy. What counted was no longer superficial differences, but resemblances in depth.

In this way a whole network of new relationships among living beings began to take outline. Until the mid eighteenth century, the character was only a fraction of the organism, an independent element selected for taxonomic convenience. The search for resemblances in performance brought the character out of isolation, to integrate it as a part of a whole. Not that the role of the character in classification disappeared: for Vicq d'Azyr or Storr, it retained the same value that Tournefort and Linnaeus had placed on it. But the character had to be considered, not separately, but in relation to the structures of the whole organism. 'Relationships are always incomplete', said Lamarck, 'when they concern only an isolated consideration, that is to say, when they are determined only after the consideration of a single part taken separately.'[21] What mattered henceforth was less the characters themselves than the relations between characters. Only the analysis of these relations allowed the constitution of a living being to be defined and its classification determined. It was no longer enough to observe in detail the resemblances and differences between the organisms. They had to be compared in the mass. 'The mind', wrote Goethe, 'must embrace the whole and by abstraction deduce a general type from it.'[22] Behind the combinative arrangement of the organs a logic of the organism was beginning to arise.

For the naturalists of that period, the different parts of a living being are no longer of equal importance for the organism as a whole. To live, a plant or an animal has first to feed and to reproduce. All component units of the organism are arranged in terms of these primary needs. If characters have different values, that is no longer the result of an arbitrary gradation, established for the sole purpose of classification, but a reflection of the relative importance of the organs in the whole structure. There are different orders of characters of unequal weight, depending on their constancy from one organism to another. For it is not enough to count characters, they also have to be 'weighed'. For instance, a character of the first order is equivalent to several of the second order, and so on. 'In adding up characters,' said Antoine Laurent de Jussieu, 'they should not be counted as units, but each according to its relative value, so that one single constant character is equivalent to, or even more important than, several ones combined.'[23] It is its role in the structure of the organism that gives weight to the character and indicates its place in the hierarchy. Organs of fructification do not occupy an important place merely because they are taxonomically useful. They are taxonomically useful because they exercise an important function – reproduction – and, by that very fact, reflect the structure of the whole plant. 'Special attention should be paid to organs of fructification,' said Lamarck in *La Flore française*,

that is, the fruit, the flower and their appendages. The principle is based in the first place on the pre-eminence naturally attached to these organs that contain the pledge of future generation, and to which is related, as if to its centre, the subordinate mechanism of the other parts which seem to live only for themselves.[24]

Subordination among characters reflects a hierarchy of structures.

Thus at the end of the eighteenth century there was a change in the relations between the exterior and the interior, between the surface and the depth, and between organs and functions of a living being. What became accessible to comparative investigation was a system of relationships in the depth of a living organism, designed to make it function. Behind the visible forms could be glimpsed the profile of a secret architecture imposed by the necessity of living. This second-order structure was organization, which brought together into one coherent whole both what was seen and what was hidden. For Lamarck, 'of all considerations, organization is the most essential guide to the methodical and natural distribution of animals'.[25] It directs investigation, for 'in animals the principal relationships will always be determined according to the internal organization'.[26] It enables one to scan the living world and to bring some order into its complexity, since 'each class should comprise animals distinguished by a particular system of organization'.[27] By this means, structural differences and functional constancy could be displayed and coordinated in the same frame. It was organization that gave living beings the internal law determining the very possibility of their existence.

The establishment of the concept of organization at the heart of the living world had several consequences. The first was the concept of the organism as a whole, henceforth viewed as an integrated ensemble of functions and, therefore, of organs. What had to be considered in an organism was not each separate part, but the whole, 'the composition of each organization in its entirety, that is, in its generality', said Lamarck.[28] If unequal values and importance could be ascribed to parts, it was always in reference to the whole. That was most clearly shown in the simplest forms of organization. 'It is particularly among insects,' stated Lamarck, 'that one observes that the organs essential to the maintenance of life are distributed almost evenly, for the most part situated over the whole body, instead of being isolated in particular places, as is the case in the most perfect animals.'[29]

Secondly, the concept of organization led to a development already glimpsed in the eighteenth century: the idea that a living being is not an isolated structure in empty space, but integrated with nature and attached to it by many different relationships. For an organism to live, breathe and feed itself, a certain harmony had to exist between the organs responsible for those functions and the external conditions. The organism had to react to what Lamarck called 'circumstances'. By 'circumstances' he meant the habitats on earth or in water, the soil, climate and the other living forms surrounding the organisms, in short, the whole 'diversity for the environment in which they live'.

Finally, the concept of organization introduced a radical division among the objects in the world. Until that time, nature had traditionally been divided into three kingdoms: animal, vegetable and mineral. In this division, things were, so to speak, on the same footing as beings, a notion justified by the imperceptible transitions then considered to exist between mineral and vegetable as well as between vegetable and animal. At the end of the eighteenth century, Pallas, Lamarck, Vicq d'Azyr, de Jussieu and Goethe rearranged the products of 'nature' into two, rather than three, groups, distinguished by the criterion of organization alone. As early as 1778 Lamarck wrote:

One will first remark a large number of bodies composed of raw, dead material which increases by the juxtaposition of the substances forming it and not because of any internal principle of development. These beings

are generally called *inorganic* or *mineral beings* ... Other beings are provided with organs appropriate for different functions and are blessed with
a very marked vital principle and the faculty of reproducing their like.
They are comprised in the general denomination of *organic beings*.[30]

From that time onwards, there were only two classes of bodies: the
inorganic was non-living, inanimate, inert; the organic was what
breathes, feeds and reproduces; it was what lives and is 'inevitably
doomed to die'.[31] Organization became identified with the living.
Beings were definitively separated from things.

Once living beings were isolated from other objects and united
through their organization, the problems of the genesis of the living
world and of the inanimate world became distinct. As Lamarck
proposed, it was no longer necessary to assume that all or most living
forms had been created simultaneously in their full complexity; they
could be derived one from another by a series of successive variations.
As the progressive tendency of nature had a cumulative effect on the
structure of organisms, the continuous series of organisms in space
could then be envisaged as the result of a continuous series of transformations in time. The emergence of living organisms, as well as
their diversity, then rested on a characteristic attribute of living
systems: their ability to vary and to adapt.

Gradually a new science emerged. It no longer studied plants and
animals as particular classes of natural bodies, but rather the living
organism endowed with singular properties as a result of a special
kind of organization. Almost simultaneously, Lamarck, Treviranus
and Oken used the term 'biology' to define this new science. 'Everything that is generally common to plants and animals,' said Lamarck,

and all faculties proper to each of these beings without exception must
constitute the unique and vast subject of *Biology*: for the two kinds of
beings which I have just referred to are all essentially living bodies and
they are the only beings of that kind which exist on our globe. The considerations pertaining to Biology are therefore quite independent of the
differences that plants and animals may show in their nature, their state
and the faculties peculiar to certain of them.[32]

Thus provided with a name and an objective, the new science progressively developed its own concepts and techniques during the

nineteenth century. Beyond the differences of form, properties and habitat, it was necessary to discover the common characteristics of living systems and to give a content to what was henceforth called 'life'.

Life

In the rationalism of the Classical period, knowledge was based on agreement between object and subject, on harmony between things and the picture the mind formed of them. With the appearance of what Kant called a 'transcendental field', at the end of the eighteenth century, the subject played an increasing role in the study of nature. Pre-established harmony was replaced by predominance of the faculty of knowing over the objects to be known. To decipher nature and find its laws, it was no longer sufficient to search for and arrange the similarities and differences between beings and things, in order to place them within a two-dimensional system of classification. Empirical data had to be brought together in depth and connected at different levels in accordance with their relation to some unifying element that was both a condition of all knowledge and yet beyond knowledge. As in every empirical field, internal analysis alone was no longer sufficient to account for the living world. It was life itself that served as a reference, a transcendental that allowed consciousness to connect the representations and establish relations, not only between the different organisms, but also between the different elements of a single organism. In the study of the living world, the concept of life made it possible to reach *a posteriori* truths and achieve a synthesis.

The very idea of organization, hereafter implicit in the definition of a living organism, is inconceivable without the postulate of a goal identified with life: a goal no longer imposed from without, but which has its origin in the organization itself. It is the notion of organization, of wholeness, which makes finality necessary, to the degree that structure is inseparable from its purpose. According to Kant, if one sees a geometrical figure drawn on the sand one can be certain that the parts of the figure are not there by chance.[33] The parts appear to be joined by an external relationship, but it is the whole structure that makes their cohesion possible, that displays order amid disorder. Exposed to the elements, the sand tends to become level and the drawing is obliterated. The figure can be formed and maintained only by an internal force that counteracts chance and destruction. An organized product of nature is both end and means. 'Each being,' said Goethe, 'contains the reason for its existence within itself; all the parts react on one another ... so each animal is physiologically perfect.'[34] Finality in the living world thus originates from the idea of organism, because the parts have to produce each other, because they have to associate to form the whole, because, as Kant said, living beings must be 'self-organized'.[35] Kant then redeveloped, in slightly modified form, the argument of the watch already used by Fontenelle. In a watch, each part is the instrument of movement of the other parts, but a wheel is never the efficient cause that produces another wheel. A part exists *for* another but not *by* another. The cause that produces wheels is not to be found in the nature of wheels, but outside them, in a being capable of putting his ideas into effect. A watch cannot produce parts that are taken away from it, replace faulty parts by other parts, or rectify the movement itself when it is out of order. An organized being is, therefore, not simply a machine, since a machine has only a force of movement, while the organism contains in itself a force of formation and regulation that it communicates to the material of which it is made.

As long as the Classical period was primarily concerned with demonstrating the unity of the universe, living beings had to conform to the laws of mechanics governing inanimate objects. The forces animating organized bodies were characterized in the terms used to describe the movement continually occurring in their fluids or solid parts. The concept of life did not exist, as shown by the definition in the *Grande Encyclopédie*, an almost self-evident truth: life 'is the opposite of death'. At the beginning of the nineteenth century, on the other hand, what mattered was to define the properties of living organisms. The study of beings could no longer be treated as an extension of the science of things. Analysis of living organisms required new methods and concepts, as well as a special language; for in the science of organized bodies words conveyed ideas that belonged to physical sciences and did not agree with biological phenomena. 'If men had cultivated physiology before physics,' said Bichat,

I am convinced that they would have applied much of the former to the latter; they would have seen the rivers flowing because of the tonic effect of their banks, crystals uniting because of the excitation of their mutual sensitivity, the plants moving because they irritated each other at great distances.[36]

In the nineteenth century, it became quite improper to describe the functioning of organisms in terms of weight, affinity and movement. To maintain the cohesion of the organism, to ensure order in the living being as opposed to the disorder of inanimate matter, there had to be a particular quality, which Kant called an 'internal principle of action'; there had to be *life*.

For the young science of biology, the idea of organization includes both what makes life possible and what is determined by it. But although life is the source of every being, it cannot be apprehended through the analysis of biological properties and functions. It is the obscure force that gives organized bodies their essential characteristics and keeps their molecules together despite the external forces that tend to separate them. As Cuvier said, it is the force that gives the body of a young woman

those rounded, voluptuous curves, that suppleness of movement, that gentle warmth, those cheeks stained with the roses of voluptuousness, those eyes gleaming with the spark of love or the fire of genius, that physiognomy enlivened by shafts of wit or animated by the fire of passions. A single instant is sufficient to destroy this magic.[37]

The living body is subject to the action of various influences, that come from things as well as beings and tend to destroy it. To resist that action, a principle of reaction is needed. Life is nothing other than this principle of struggle against destruction. For Bichat, life is 'the sum of the functions that oppose death';[38] for Cuvier, 'the force that resists laws governing inanimate bodies';[39] for Goethe, 'the productive force against action of external elements';[40] for Liebig, 'the motor force that neutralizes the chemical forces, cohesion and affinity acting between molecules'.[41] Death is the defeat of that principle of resistance and a corpse is nothing more than the living body that has been reconquered by physical forces. The powers of order, unification and life are constantly battling against those of disorder, destruction and death. The living body is the theatre of that fight, in which health and disease reflect the changes of fortune. If the vital properties win, the living being regains harmony and recovers. On the contrary, if physical properties are stronger, death results. Nothing like that is to be seen in inanimate bodies; things are immutable, like death.

However, although originally the organization, functioning and entire architecture of the living being required the intervention of a principle of life, in the end the living being becomes engulfed in life. Whereas physical properties of matter are eternal, the living properties of an organism are fleeting. Inanimate matter enters living bodies to become steeped in vital properties. The organism becomes 'a sort of furnace', said Cuvier, 'to which dead substances are transported successively, there to combine together . . . and to escape one day and once more become subject to the laws of dead nature.'[42] During the life of an organism, physical properties are, so to speak, 'fettered' by vital properties; consequently, they are prevented from producing the phenomena that their nature leads them to produce. But it is not a lasting alliance, for it is a characteristic of vital properties to become rapidly exhausted. 'Time wears them away', said Bichat.[43] The fact that it is alive means that the organism is doomed to die. The living being captures the power of life in some way; it fixes and immobilizes this force, but only momentarily, since it is destroyed precisely by what has made life itself spring forth. 'If life is the mother of death,' said Cabanis, 'death in its turn gives birth to life and perpetuates it.'[44] The living is reduced to a heap of matter that life has fleetingly touched. Although the vital properties are worn away in each organism, they are preserved in the living world. Whether born of a seed or a foetus, each living body has once been part of a similar body. Before acquiring autonomy, before becoming in its turn the seat of an independent life, each organism has first shared the life of another organism from which it has subsequently become detached. Life is transmitted from being to being in an unbroken succession. Life is continuous.

This vitalism was quite different from the animism of the Classical period. In the nineteenth century, the reference to a vital principle stemmed from the attitude of biology itself, from the need to separate beings from things and to base that separation, not on matter whose unity was recognized, but on forces. Vitalism operated as a factor of abstraction. Life played a precise role in knowledge. Life was the very object of investigation, what was studied in the animal or the plant. It was that part of the unknown which made the living being different from an inanimate object and the science of biology distinct from that of physics. Vitalism was as necessary for the establishment of biology as mechanism had been for the Classical period – necessary not only for naturalists, physiologists and physicians, but also for chemists studying 'organic compounds', the substances that composed living beings or were produced by them. 'One may consider,' said Liebig, 'that the reactions of simple bodies and mineral compounds prepared in our laboratories can find no kind of application in the study of the living organism.'[45]

The Chemistry of Life

At the end of the eighteenth century, the composition of various organic bodies was already being studied. In particular, Scheele and Bergman had analysed a whole series of organic acids and isolated the 'sweet principle of oils', later to be known as glycerol. At the beginning of the nineteenth century, methods of analysis were improved, particularly with the discovery of electrolysis. The atomic theory became more clearly defined. An immense range of compounds was included under the heading 'organic chemistry'; they all contained carbon and hydrogen, often oxygen and sometimes nitrogen, sulphur or phosphorus. These compounds were either the constituents of living beings or their products of excretion and decomposition. Lavoisier had introduced a method of classification and a universal nomenclature in chemistry. Berzelius created a universal code: each element was represented by the first letter of its Latin name, followed, if necessary, by another letter to prevent confusion. The formula of a compound was established by juxtaposition of the symbols representing the constituent elements, each symbol being given a numerical coefficient showing the number of atoms in the molecule. Improved analytical methods and the use of these symbols made it possible to manipulate and symbolize the enormous molecules which composed living beings.

To establish organic chemistry, however, it was not enough to improve the techniques of mineral chemistry. Also necessary was the outlook of biology which, seeking beyond the diversity of organisms, was beginning to define the unity of the living world. As long as that world was envisaged as containing an infinite number of structures, an infinite number of compounds could be expected. If, on the other hand, living systems were characterized by a special organization and special functions, if they were always what feeds, grows and multiplies, then it became possible to study the nature of the compounds that distinguish living beings from non-living matter, as well as the reactions by which living beings transform food-stuffs and incorporate them. This delimited a field which verged on chemistry on one hand, since the substances of living beings were formed of the universal chemical elements that compose all matter, and approached biology on the other, since these compounds were radically different from those studied in inorganic chemistry.

What was most directly accessible to chemical investigation was the flow of material through living beings – the 'metamorphosis' of food into characteristic compounds and the formation and excretion of waste products. It was, therefore, the task of organic chemistry to study the transformations of substance within the organisms and to recognize the nature of the elements and their compounds as they enter and leave the organisms. But these transformations appeared as a challenge to the laws of ordinary chemistry: when united in new combinations, the constituent elements of living beings exhibited totally different properties. Organic chemistry had to identify the substances that make up living organisms and pass through them. It had to try to analyse them. But it was not expected to furnish proof of these transformations by chemical synthesis. 'All the actions of the [body's] economy', said Liebig, 'are subordinated to some non-material activity which the chemist cannot use at will.'[46] Wöhler performed a preparation of urea in the laboratory by boiling a solution of ammonium cyanate – but he himself refused to consider it as the synthesis of an organic compound from inorganic substances. For although the raw material, hydrocyanic acid, was a relatively simple compound, it was still an organic substance. 'A philosopher of nature', wrote Wöhler to Berzelius, 'would say the organic animal character has not disappeared from the animal carbon and from these cyanate combinations, and for that reason one can obtain other organic bodies.'[47] It was not until Berthelot produced acetylene from carbon and hydrogen that the barrier interposed by the chemists between organic and inorganic was to be torn down. Yet, by means of analysis alone, scientists, in the first half of the nineteenth century, revealed the presence of a considerable number of different compounds in living beings, some containing nitrogen and others not. Depending on their composition, these substances played different roles in the organism.

Combinations of the same elements were often found to exhibit different properties, depending on whether they occurred in mineral or organic substances. Living beings had, therefore, to contain a particular force which brings about a change of form and motion in matter, which disturbs and destroys the state of chemical inactivity that maintains the combinations of chemical elements in foodstuffs: it was 'vital force'. 'It causes decomposition of foodstuffs,' said Liebig,

it upsets the attractions which continuously affect their particles; it diverts the direction of chemical forces so as to group round itself the chemical elements of foodstuffs and produce new compounds, . . . it destroys the cohesion of foodstuffs and makes the new products unite in new forms distinct from those they have when the force of cohesion acts freely.[48]

The chemical forces assembling the atoms in inorganic molecules act in living beings as a resistance, to be overcome by the vital force. Were the two forces of equal strength, there would be no effect, no growth, no reproduction. Were the chemical force to win, the organism would perish. For an organism to live, the vital force must carry off the victory. Vitality cannot be attributed to any particular organ, tissue or molecule. It is a property of the complete organism, a characteristic of the whole that results, said Liebig, from 'the union of certain molecules in certain forms'.[49] It depends on the very organization of living beings.

If vital force became a concept of such importance at the beginning of the nineteenth century, it was because it then played a role subsequently assumed by two new concepts. Today, living organisms are seen as the site of a triple flow of matter, energy and information. In its early days, biology was able to recognize the flow of matter; but, lacking the other two concepts, it had to postulate a special force. Indeed, until the middle of the nineteenth century, the relation between heat and work remained very vague. Carnot associated heat with the movement of the corpuscles that constitute bodies. Although the work of Carnot was later to become a kind of birth certificate of the second principle of thermodynamics, it was ignored for nearly twenty years, until the moment when the law of equivalence and the concept of energy made it possible to integrate all phenomena in which heat plays a role. Meanwhile, a factor had to be found to neutralize the forces of affinity between elements in molecules, to redistribute these elements through different chemical bonds and to regroup the atoms in new compounds. Some force had to be invoked which, together with sunlight, separated oxygen in plants from the elements for which it had the most affinity and released it in the form of a gas. 'A certain quantity of vital force,' wrote Liebig, 'must be spent to keep the elements of the nitrogenous principles in the order, form and composition which characterize them, and also to resist the incessant action on their elements of atmospheric oxygen and oxygen secreted in plant growth.'[50] This is exactly what modern biochemists say, but 'vital force' is replaced by 'energy'. For Berzelius, Liebig, Wöhler and Dumas, 'vitality' was not a principle acting at a distance, like gravity or magnetism, but a force that takes effect 'within an agglomeration of material', when the substances of a reaction came into contact. To manifest itself, vitality required a certain degree of heat, since all the phenomena of life ceased when an organism was exposed to cold. It was the combustion of atmospheric oxygen with certain food substances that provided the heat. The substances that could be oxidized in this way and play a 'respiratory role' were mainly non-nitrogenous compounds, such as sugars and fats.

Nitrogenous compounds, in contrast, played a 'plastic role' in the make-up of organs and tissues. But in order to explain their composition and production it was once again necessary to refer to the vital force. In all living beings and tissues studied, very complex but similar nitrogenous compounds were found. When these compounds were analysed – whether fibrin and albumin in blood, or casein in milk – they always revealed a mixture of carbon, hydrogen, oxygen and nitrogen in rigorously constant proportions, with, in addition, variable quantities of other elements, notably sulphur and phosphorus. It had to be admitted, therefore, that all living tissues were composed of the same basic constituent capable of combining with different quantities of other elements. Consequently, it was those particular combinations that gave the different organs and tissues their characteristics. Mulder called this basic constituent 'protein' to emphasize its primary character. It was through the production of protein and its combination with certain elements that the architecture of living beings was built. 'It must be admitted as a law demonstrated by experiment,' wrote Liebig, 'that plants elaborate proteinaceous compounds, and that it is these compounds that are moulded by the vital force, under the influence of atmospheric oxygen and the components of water, to create all the numerous tissues and organs of the animal economy.'[51] However, if substances of similar composition could possess distinct properties, a new principle had to be invoked. It had to be admitted that, in such molecules, the same atoms could occupy different places and that it was the position of the atoms that determined the nature and properties of the molecules. Analysis of a series of simpler compounds led to the same idea. Pairs of substances as different in their properties as cyanates and fulminates, or racemic and tartaric acids, were shown by analysis to possess identical elements. There again the position of the atoms in the molecule had to be the explanation. 'Isomerism' was the name Berzelius gave to this difference in structure coupled with identity in composition. There exists then a principle of order of molecular structure, according to which the nature and properties of a molecule depend on the relative position of its atoms. In modern biology, this molecular order, this choice between possible structures, is interpreted by the concepts of entropy and information. In organic chemistry at the beginning of the nineteenth century, a mysterious force had to be invoked to assign atoms to their places.

As for the mechanisms of the reactions occurring in living beings, they were obviously different from those produced by the chemist in the laboratory. Respiration is certainly combustion, as Lavoisier had shown, but it is combustion of a very special kind. In organisms, foodstuffs are consumed very slowly at a low temperature, not rapidly at a high temperature as in an oven. In the laboratory, sugar can be carbonized, but not transformed into ethyl alcohol and carbon dioxide as it is by brewer's yeast, or into butyric acid and hydrogen as by reaction in cheese. In organisms, therefore, there exist certain principles or substances, called 'ferments' or 'diastases', that direct the course of the chemical reactions to alter the bonds between elements and thus change a substance into new products. There are two ways of looking at these ferments. First, like Liebig, one can consider that certain substances are able to transmit some of their properties to other substances. Ferments are then substances whose atoms are in a state of great agitation and that transform other compounds by transmitting their own motion to them.

The phenomena of decomposition can only be explained by accepting that they are the result of contact with a substance which is itself in a state of decomposition or combustion ... The motion of the molecules of one of the reacting substances must have a certain influence on the equilibrium of the molecules of the substance in contact with it.[52]

Brewer's yeast ferments sugar because the yeast contains a substance in a state of 'metamorphosis'. By virtue of a chemical quality peculiar to ferments, the decomposition spreads beyond the sphere of that substance and affects the molecules of a neighbouring compound.

But like Berzelius, one can also relate the properties of ferments to a new chemical force apparent in the transformation of certain compounds, both mineral and organic. In solid state or in solution, many simple or compound bodies prove able to counteract the chemical

forces holding other compounds together and to cause their transformation. These substances act in a very special way. They modify
the relations existing between the atoms of one substance, without
themselves playing a chemical role in the reaction, since they are
found intact at the end of it. In this way manganese, silver or blood
fibrin cause the decomposition of hydrogen peroxide; sulphuric acid,
like the diastase from germinating seeds, changes starch into sugar;
and finely divided platinum at ordinary temperature can either ignite
pure alcohol, or oxidize diluted alcohol to acetic acid. Berzelius gave
the name 'catalysis' to this type of reaction of unknown origin, that
takes place in the presence of certain bodies.

> The catalytic force consists in the fact that certain bodies can by their
> simple presence...awaken chemical affinities which otherwise would remain
> inactive at the given temperature... Consequently, it acts rather like heat.[53]

The reactions that take place in living beings can accordingly be
compared to catalytic reactions. Diastase, for instance, is not found in
all parts of the potato, but only in the 'eyes', where the starch is
changed into dextrin and sugar. Because of the catalytic reaction, the
part round each eye becomes a centre for the production of juices
that provide nourishment for the young shoots. 'It is probable,'
said Berzelius, 'that in the plant or in the living animal, thousands of
different catalytic processes take place, whereby the uniform raw
material of plant juice or blood gives rise to a quantity of different
chemical compounds ...'

Accordingly, adopting ideas and techniques derived from chemistry, an empirical science was set up which investigated the composition of living beings and gradually fashioned its own concepts and
vocabulary. Little by little, the nature of the fabulous compounds
that only organisms seemed able to form was defined. Once again,
behind the complexity of molecular architectures the simplicity of a
combinative system became outlined. This was the case, for example,
of fats and oils, which had always been a favourite object of study for
chemists. From the days of alchemy to the eighteenth century, lard,
suet and butter had constantly been beaten, kneaded and burned,
without their nature or composition ever being defined. By using less

crude methods of analysis, Chevreul found that fats were produced by an arrangement of several simpler compounds. Just as alloys were formed by amalgamating certain metals in certain proportions, so fats were made up by the combination of a 'sweet principle' – the future glycerol – with certain so-called 'fatty' acids. There exists a great variety of these fatty acids. It is the type of acid combined with glycerol that determines the nature and properties of fats and oils.

In the reactions involving organic compounds, it happens that one element takes the place of another, dislodging it in a way, without the architecture of the molecule being destroyed. These 'substitutions' may affect not only isolated elements, but also groups of atoms, 'radicals' that remain linked together during chemical transformations. Radicals can also be added to a molecule as a unit, separate from it again, and receive extra atoms without themselves ever being changed. With the work of Dumas, Laurent, Gerhard, Liebig and Wöhler appeared families of compounds, 'types' or 'nuclei', on which a variety of radicals could be hooked to give the molecule certain chemical functions: alcohol, aldehyde, acid, amine, ether, etc. A double-entry classification of organic compounds was thus established. On the one hand there were homologous series which enabled the substances to be classified in natural families according to their composition. On the other, there were chemical functions that established relationships between the compounds of different families with similar properties.

Finally, the immense diversity of organic compounds could be reduced to a combinative system of a limited number of types and functions. The variety of molecules and of their properties was born of the motion of certain atoms or groups of atoms that can come and go without the entire architecture being changed, somewhat like the framework of a building where stones or tiles may be substituted without affecting the basic structure. Behind the diversity of living forms, of organs and of compounds, there began to appear the operation of chemical reactions that attack foodstuffs and rearrange them, in order to construct the types of molecules necessary for life and to excrete the waste products. Placed at the crossroads of biology and chemistry, the new science sought to define the exact limits of life which at the beginning of the nineteenth century was claimed to be specific and irreducible. Organic chemistry measured the distance between things and beings, between what was amenable to the laws of physics and what was not. It was precisely this distance which was destined to be reduced at the end of the century, thanks to two new ways of conceiving order in matter: the order that statistical mechanics was to derive from the disorder of molecules; and the order that physical chemistry was to insert into the structure of molecules.

The Plan of Organization

At the beginning of the nineteenth century, naturalists were attempting to identify the order that reigned, not only among living beings, but also in the innermost recesses of the organism itself. Animals rather than plants then became the chief objects of investigation. Although complexities of structure are more conspicuous in plants, the demands of organization are more obvious in animals. Behind the complex architecture of an animal, the mystery of its functions can be divined. Everything there combines to produce the incessant quivering characteristic of life. The behaviour of animals, their passage from health to disease, the dangers that everywhere lie in wait for them, all bear witness to the constant battle between the forces of life and the forces of death.

To study the organization of an animal, it is not enough to dissect it, to distinguish all its parts and map them. Organs have to be examined in terms of the role they play in the whole organism. But the methods of chemistry are still forbidden in physiology. Separating parts of the body to study them means denaturing them, for, wrote Cuvier, 'the machines that are the subject of our researches cannot be dismantled without being destroyed'.[54] Details of morphology pale in comparison with the complete living being. The arrangement of the anatomical parts reflects an internal bond, a coordination of functions that links the structures in depth. While the function is a fundamental requirement of life, the organ is only a means of execution. Although the function brooks no fantasy, the organ preserves some slight degree of freedom. In surveying the whole animal kingdom, it is therefore possible to see what is constant and what is variable, and thereby determine the degree of variation in the organ which the function tolerates. Consequently, wrote Cuvier, living bodies are 'kinds of experiments prepared by nature, which adds or subtracts parts from each, as we might like to do in our laboratories, and shows us the result of these additions or subtractions'.[55] Behind the different forms the community of functions has to be revealed. The difference in structure between a leg and a wing is less important than the similarity of their roles. The lung and the gill might well look completely different; nevertheless, they are both organs for breathing, either in air or in water. Morphological differences between testicle and ovary, epididymis and fallopian tube, penis and clitoris, should not hide the symmetry of the two series, the similarity of roles and anatomical connections. Irrespective of the organism, the phenomena of life can occur only within the shelter of an envelope which protects it from external elements. 'No matter whether the envelope be skin, bark or shell,' said Goethe, 'everything that is living, everything that acts as if endowed with life, is provided with an envelope.'[56]

Those resemblances, based on the criterion of position and function and no longer of shape, revived the old Aristotelian concept of analogy. Naturalists admitted that structures could vary in shape in different species, depending on the role they played. According to Geoffroy Saint-Hilaire, one can

trace the forefoot just as well in its various uses as in its numerous metamorphoses, and see it utilized successively in flight, swimming, jumping, running, etc.; here it is a digging tool, there a hook for climbing, elsewhere an offensive or defensive weapon; it can even become, as in our own species, the principal organ of touch and, consequently, one of the most effective instruments of our intellectual faculties.[57]

For Geoffroy Saint-Hilaire and Cuvier, in fact, the word 'analogy' covers two different aspects later distinguished by Owen under the terms 'homology' and 'analogy'. Homology describes the correspondence of structures, analogy that of functions. Homologous organs occupy the same position and play a similar role in different species: for instance, a man's hand and a bird's wing. Analogous organs fulfill the same functions in different species, despite anatomical differences in structure, position and relation: for instance, the viscera of digestion, or the liver found in different forms in crustacea, molluscs and vertebrates. Comparison of animals of the same class shows the existence of certain relations between the structure, position and function of the respective organs, despite innumerable differences in size, shape and colour. Variations in shape are not distributed at random. Each component is connected to the others to ensure the harmony of the whole. It is 'the mutual dependence of functions', said Cuvier, 'and their reciprocal aid that are the basis of laws determining the relations between their organs, laws which are as necessary as metaphysical or mathematical laws'.[58] The object of analysis is no longer a random collection of structures in an infinite number of combinations, but a system of relationships that dovetails in the innermost part of the organism. To analyse and, indeed, to classify living beings, the combinations have to be arranged round the most important functions, such as circulation, breathing, digestion and so on. The real aim of zoology becomes the study of the different ways of carrying out these functions; its principal tool, comparative anatomy.

For the nineteenth-century biologists, there are two approaches to comparative anatomy. The first involves almost exclusively the study of morphology. Reference to physiology is made by limiting the analysis to large functional sectors, called 'regions' by Geoffroy Saint-Hilaire. Consequently, in a whole series of species, corresponding structures or 'analogies' have to be looked for, region by region. The gill covers in fish are compared with the auditory ossicles of the ear of air-breathing vertebrates, for instance; or parts of the larynx, trachea and bronchi of land animals are compared with the arches, teeth and cartilaginous laminae of the gills in aquatic animals; or, again, the composition, shape, position and anatomical connections of the hyoid bone are compared in fish, birds and mammals. Then, using as point of reference the species in which the region in question is maximally developed, an attempt is made to arrange other morphological types in series based on the displacements and deformations which they display. Very often a single organ in different species shows a series of gradual transitions between two extremes. Ultimately, despite modifications, the same elements of structure and the same number of elements are discovered. The homology then becomes obvious.

Sometimes, however, it is impossible to form a series, either because the transitional forms are still unknown, or because they have disappeared. In that case, to identify a part and recognize analogies, it is necessary to resort to 'connections'. For the most constant characteristic of a region lies in the relations between its parts. Whatever changes in shape, volume or position an anatomical part may undergo, it always retains the same relations with the surrounding area and remains linked to the same surrounding components. 'An organ,' said Geoffroy Saint-Hilaire, 'is altered, atrophied or abolished rather than transposed.'[59] The principle of connections therefore makes it possible to identify a part which appears in another species. It is generally an unusual development of a formation that also exists elsewhere. Within a given region, different structures are not independent. There is a 'balance of organs' so that the excessive development of one part affects its neighbouring parts. A normal component never acquires a new property without another component of its system or a related part inevitably being affected. 'If an organ happens to expand enormously,' wrote Geoffroy Saint-Hilaire, 'its influence becomes tangible in the neighbouring parts which thereafter no longer attain their normal development; all these parts are nonetheless preserved.'[60] When the shoulder bones of vertebrates are examined in terms of their role in breathing, variations in shape, size and position are observed that make it possible to distinguish four degrees of development: the two extremes are hypertrophy and the rudimentary state. In fish, the shoulder bones pass above the heart and behind the gills, to act as a sternum. A particular structure can therefore be integrated into neighbouring organs, either retaining its function or acquiring another. 'The general total in nature's budget is fixed,' wrote Goethe; 'but she is free to allocate partial sums for any expense she wishes. If she spends in one way, she is obliged to economize in another, and that is why nature can never fall into debt or become insolvent.'[61]

Finally, if even the search for connections does not permit the recognition of homologies, it becomes necessary to examine the arrangement of the regions in the embryo. Anatomical peculiarities frequently appear during embryonic development, only to disappear later in the adult. By this approach, the bony parts forming the skull of different vertebrates can be identified. In adults, the skull of fish seems to be made up of a greater number of parts than that of mammals. But this difference disappears when the embryonic skulls are examined and the number of bones counted by the number of centres of ossification. It is then seen, said Geoffroy Saint-Hilaire, 'that the skull of all vertebrate animals comprises about the same number of parts, and that these parts always keep the same arrangement and the same relationships and are used for similar purposes'.[62]

Comparative anatomy can also involve a closer association between physiology and morphology. The organisms are then viewed as a whole, not region by region. Variations in structures are compared only to detect the permanency of functions. According to Cuvier, anatomy becomes a tool for reaching 'the laws of the organization of animals and the modifications this organization undergoes in the different species'.[63] The very existence of an organism depends not only on the execution of certain functions, but on their coordination. The living body, therefore, cannot be a mere collection of organs combined in various ways for performing those functions. The organs must also be arranged to produce a harmonious whole, 'because in the living state,' said Cuvier, 'organs are not merely close together, but act on each other and all cooperate for a common aim ... There is no function that does not require the aid and cooperation of almost all the others.'[64] Consequently, any modification in one structure exercises an influence on the others. Certain variations are mutually exclusive. Others seem, so to speak, to attract one another, not only between neighbouring organs, but also between those which at first sight appear most distant and hence most independent. From the association between functions can be deduced the 'law of coexistence' which determines the relationship between organs.

An animal that digests only flesh must be able to see its prey, follow it, overcome it and tear it apart. Consequently it must have a piercing eye, a keen sense of smell, a swift gait, agility and strength of leg and jaw. For that reason, cutting teeth for tearing through flesh are never found in the same species with a foot cased in horn that can only support the weight of the animal and cannot be used for grasping.[65]

Hence the rule by which every hoofed animal is herbivorous. Hooves indicate flat-crowned molar teeth and a long alimentary tract. The laws that determine the relationships between organs with different functions apply equally to the different parts of the same functional 'system' to coordinate the variations. In the digestive system, for instance, the shape of the teeth, the length, coils and dilatations of the alimentary tract, the number and quantity of digestive juices always show a particular relationship.

The law of coexistence and the correlations that it permits changed the way in which living organisms were observed and studied. Cuvier was no longer concerned with detecting the components in order to establish their variations by comparison with other organisms. Structures were arranged in layers in the depth of the body in accordance with a secret rule which had somehow to be discovered by analogy. A living organism was a 'unique and self-contained' whole. All its parts were mutually related and cooperated for the same purpose. Although no part could change unless the others also changed, each part considered separately sufficed to indicate the others. 'There is hardly any bone', said Cuvier, 'that varies in its facets, curves and protuberances without the others undergoing corresponding differences; consequently by looking at one single bone, the appearance of the whole skeleton can be deduced up to a certain point.'[66] Palaeontology was founded on this principle of reconstructing extinct organisms from the few remaining fossilized fragments. In comparative anatomy, a rediscovered fragment was no longer an isolated part. It was the token of a whole organization.

However, organs were not considered to be connected merely by a network of correlations. They were part of a hierarchy imposed by the very existence of the living being. Antoine Laurent de Jussieu had already shown that characters were subordinated; but that was still founded on a structural criterion. If certain characters were found more often than others, they ought to be more important. For Cuvier, the importance of the character is only the measure of its functional importance, the subordination of structures refers to a hierarchy of functions, a coordinated system that controls the distribution of organs. The relative importance of an organ can be estimated by the constraints it imposes on other organs. Certain structural features exclude, or alternatively demand, certain other features. The relations between different organs can thus be 'calculated'. 'The parts, properties or structural features,' said Cuvier, 'that have the greatest number of relations of incompatibility or coexistence – that is to say, that exercise the most marked influence on the whole being – are the *important* or *dominant* characters. The others are subordinate characters.'[67] This provides a means of recognizing important characters: in a series of organisms, they are the most constant. If organisms are compared in terms of their similarities, the important characters are those that vary least. This is true equally of 'animal functions', such as sensitivity and voluntary movement possessed by animals alone, and of 'vegetative functions', such as nutrition and generation that are common to animals and plants. The heart and circulatory organs form a centre for vegetative functions just as do the brain and the trunk of the nervous system for animal functions. A survey of all animals shows that these two systems are gradually degraded until they both disappear. 'In the lowest animals, where no more nerves are visible, there are no distinct fibres and the digestive organs are simply hollowed from the homogenous mass of the body. The vascular system disappears even before the nervous system in insects.'[68] The classification of organisms could, therefore, no longer be founded on structural criteria alone. Their organization of functions underlay classes, drawing certain organisms together and setting others apart. The correlation of forms that resulted from the arrangement of the motor organs, from the distribution of the nervous material and the extent of the respiratory system, was indeed the basis on which divisions had to be made in the animal kingdom.

For the nineteenth century, the very existence of a living organism depended, therefore, on harmony between its organs, which itself derived from the interaction of its functions. This consequently changed the limits of the possible for living organisms. For the eighteenth century, all observable differences in form could be combined *ad infinitum* to produce all imaginable varieties of living bodies. In the nineteenth century, such a view had only abstract value. Not all variations were permissible: only those combinations satisfying the functional requirements of life could be realized. The structure of an organism had to conform to an overall *plan of organization* co-ordinating the functional activities. However, although the findings of comparative anatomy plainly demonstrated the existence of such a plan, it did not have the same significance for Geoffroy Saint-Hilaire as for Cuvier. According to Geoffroy Saint-Hilaire, there are no anatomical structures peculiar to a species. No part appears here and disappears there. What exists in one also exists in the other, although changes may be sufficiently great that the determination of analogies becomes difficult. That is to be seen, for example, when a certain part of a region takes on such importance that it affects the neighbouring parts. As a result, the latter no longer attain their normal development. Nevertheless, they are all preserved. They can usually all be recognized, even if, reduced to their simplest expression, they have become useless 'rudiments'. For Geoffroy Saint-Hilaire,

Nature constantly uses the same materials and her only ingenuity is in varying the forms. In fact, as if she were confined to using primary data, she constantly tries to create the same number of the same components in the same circumstances and with the same connections.

As if the structure of animals were following a single plan – not one for vertebrates alone, another for molluscs and yet another for insects – but a 'general plan' for all organisms of the animal kingdom. Vertebrates and invertebrate animals, for instance, are distinguished by a change in form and not a change in constituent parts that preserve their arrangement and connections. According to Geoffroy Saint-Hilaire, it can be said that 'each part of an insect finds a similar place in vertebrate animals, that it is always in that place and that there it always remains faithful to at least one of its functions'.[70] Insects live within their spinal column as molluscs live within their shells.

It was not the first time that the idea of a plan of construction for all living beings had been put forward. Already in the second half of the eighteenth century it had often been suggested. Buffon always found 'the same basic organization' throughout the living world. Daubenton saw in it 'a primary and general design'. For Vicq d'Azyr, nature seemed to 'operate always according to a primitive and general model from which she departs reluctantly'. For Goethe, there existed an 'essential form with which nature never ceases to play'. What underlay the idea of a single plan controlling the composition of all organisms was still, up to the time of Geoffroy Saint-Hilaire, the old notion of continuity in the living world, the chain of beings seen by the eighteenth-century scientists. It was still necessary to refer to a continuity, not of visible forms, but hidden in the very depths of the living, in order to discover a single model and a single type of organization for the whole animal kingdom.

This continuity was broken by Cuvier. The plan of organization became the place where two series of variables were articulated, one outside and the other inside living bodies. 'The different parts of each being,' said Cuvier, 'must be coordinated to make possible the whole being, not only in itself, but in relation to what surrounds it.'[71] On one hand there is, according to Cuvier, the world in which the organism lives and that determines its 'conditions of existence'. The organism is not an abstract structure that lives in a void. It occupies a certain space in which it has to perform all the functions required by life. It is extended beyond itself to the earth it tramples, the air it breathes and the food it absorbs. 'Its sphere,' said Cuvier, 'reaches beyond the limits of the living body itself.' A whole network of interactions is thus established between what lives and what allows it to live. Among all the possibilities, the living being has to remain within the limits prescribed by the conditions of existence.

On the other hand, there is the organization of the body. Continuity is no longer found in shapes and structures, but in the functions which have to be coordinated to fit the conditions of existence. By means of functions, analogies are distributed throughout the living world. All found together in the higher species, they disappear one after the other when increasingly simpler forms are examined. So long as the harmony of the whole is preserved, the executive agents – that is, the organs – remain completely free to vary. In theory, each organ could thus be changed indefinitely and each variation could be combined with all the variations of the other organs to form a continuous whole. In practice, nothing of the kind occurs, since the organs are not independent elements. They react with each other. If each organ is taken separately, it is possible to follow its gradual degradation throughout the living world. It can still be recognized in vestigial form among species in which it is no longer of use, as if nature did not want to suppress it entirely. However, all organs are not degraded in the same order among animals. 'In such a way that if one wanted to classify the species according to each organ separately, there would be as many series as regulatory organs.'[72]

In the long run, a survey of the animal kingdom does not show a linear series progessing from one extremity to the other by a series of intermediates, but discontinuous masses totally isolated from each other. Although the same functions always occur, they belong to different hierarchies and are performed by different organizations. For Cuvier, there is, therefore, no single plan for the whole living world, but several.

Four principal forms exist, four principal plans, if one may say so, by which all animals seem to have been modelled, and the later divisions of which are only quite slight modifications founded on the development or the addition of some parts, which changes nothing essential in the plan.[73]

The living world is thus made up of little isolated islands separated by impassable channels. The Cephalopods, for instance, lead 'to nothing else'. They cannot have resulted from the development of other animals, and their development 'has produced nothing superior to them'. Nature is thus seen to 'leap' from one plan to another. Between her productions she leaves 'an obvious hiatus'.[74] To make it perfectly clear how impossible it is to link the large groups of animals in a continuous series, the groups are called 'branches'. There are no intermediates between the first two branches. Molluscs and vertebrates have nothing in common, no resemblance whatsoever, either in number or parts or organization. It is only within each group, by following each branch, that series can be found, and even then they are not linear.

The Cuttle-fish and the Cephalopods are so complicated that it is impossible to find any other animal that could reasonably be classified between them and the Fish, and within its class there is a series of degradations of a common plan followed as closely as among vertebrates, so that one can work down from the Cuttle-fish to the Oyster, rather like from Man to the Carp; but there is not a single line leading down one branch to another.

At last the chain was broken: the chain that had heretofore united all living beings as if there were always an infinite number of intermediates to fill the gaps between two related organisms. A breach was opened, not only between beings and things, but also between groups of beings. It was no longer possible to progress through the animal kingdom by a single series of gradations and nuances. It was no longer possible to proceed from one extremity to another, by adding a little structure, a little complexity, a little perfection. Throughout the living world the same functional requirements existed. Everything had to feed, breathe and reproduce, whether in air or water, in hot climates or cold, in the light or in darkness. Henceforth continuity was located in the functions, not in the means of performing them. To display beings and fit their organization into the conditions of existence, nature proceeded by leaps and bounds. According to Cuvier, an animal is curled round a 'centre of organization', a centre that commands the deep-seated arrangement of its structures. It is set out in concentric masses round a nucleus. The organs are arranged in layers from the centre outwards, the most important in the middle and the subordinate on the outside. The essential is thus buried in the farthest depths of the organism, while the minor parts are displayed on the surface. The heart of the organization can almost never vary since, to change it, everything else has to be changed and the plan replaced by another plan. Secondary organs, on the contrary, can vary at will. The less important they are and, consequently, the nearer the surface, the more freely they can be modified. 'When one reaches the surface, where the nature of things obviously intended to place the least essential parts that it is less dangerous to damage,' said Cuvier, 'the number of varieties becomes so considerable that all the work of naturalists has not yet been able to describe them.'[75] Identical in the centre of the organization, organisms of the same group progressively separate from each other towards the exterior. Similar in what is hidden, they diverge in what is visible. It is on the surface that small continuous series of numerous variations can be found. In depth there can be only radical changes, only leaps from one plan to another.

What was radically transformed at the beginning of the nineteenth century, was, therefore, the way in which living beings were arranged in space: not only the space in which all beings were disposed, broken into separate islands and carved into independent series – but also the space in which the organism itself took up its abode, coiled round a nucleus, formed by successive layers that extended beyond the living being, linking it to its surroundings. It was both the relations established between the parts of an organism and those uniting all living bodies that were entirely redistributed.

The Cell

In the nineteenth century, biologists were able to extend their studies of organization to another level in the structure of living beings. In addition to what might be called macro-organization – that is, what the zoologist studied when he tried to discern the plan coordinating functions behind the intermingling of the organs – a micro-organization of living beings was revealed. That was the inmost structure of organized bodies, their elementary composition that, despite the diversity of forms, gave its special quality to the substance of each being – a texture and a set of properties that inorganic bodies were lacking.

The discovery of the cell has often been attributed to the seventeenth century. However, although by studying sections of cork and certain vegetable parenchyma under the first microscopes, Robert Hooke, Malpighi, Grew and Leeuwenhoek had been able to perceive rows of alveoli that Hooke had called 'cells', there was no possible generalization or use for such structures. When Maupertuis and Buffon had tried to introduce discontinuity into the substance of living beings with their idea of constituent elements, they were only behaving like good disciples of Newton. Living particles and organic molecules were merely the means of extending to living bodies the discontinuous nature of matter and bringing the world of beings into line with that of things, according to the requirements of eighteenth-century mechanics. But in order to make the properties of organisms correspond to the structure of matter, it had been necessary to invoke molecules of a special type reserved for living bodies alone. In the end, only the special nature of their molecules distinguished the composition of living beings from that of inanimate things. In the eighteenth century, however, the component element of living bodies had been the ultimate point in anatomical analysis, what had been found by dissecting muscles, nerves or tendons: it was the fibre. Moreover, in most of the organs, the fibre had remained a product of thought visualized as a cluster of molecules linked by a viscous substance, 'gluten'. According to Haller, there was only one kind of fibre to form all organs. The same fibres were interwoven in a continuous web extending from bone to tendon, tendon to muscle, muscle to nerve and blood-vessel. It was the way the fibres were arranged, the texture of the network they formed and the quantity of liquid retained by the mesh that made an organ firm or flabby, rigid or supple.

With the nineteenth century, the situation changed. Despite the diversity of their forms, the same organs always performed the same functions. With respect to the specific structure of organs, there were no longer warm-blooded animals, mammals or reptiles, but tendons and blood-vessels, bones or membranes that always played the same role and had necessarily to be of the same nature. If there were differences between the muscle of a pigeon and that of a frog, it was not only because one occurred in a bird and the other in an amphibian. It was also the result of external circumstances: of the difference in roles and surroundings. The similarity of functions demanded unity of structure.

Conversely, with Pinel and Bichat, the various organs that played different roles in the same living being could no longer have the same composition. 'The least reflection is enough,' wrote Bichat, 'to understand that these organs must differ, not only in the way in which the fibre that forms them is arranged and interwoven, but also, in the very nature of the fibre itself; they differ in composition as in tissue.'[76] It is no longer the shape alone that gives an organ its properties: it is first and foremost the specific nature of its component tissue. At first sight, a great variety of tissues seems to exist in living bodies. However, that is only an impression, for the tissue is characteristic, not of the organ, but of the 'system', whether nervous, vascular, muscular, osseous or ligamentary. The system represents the point of juncture between anatomy and physiology, thanks to the quality of tissue from which it is formed. A living body is thus filled with layers of tissues, sheets of membranes that cover several organs and divide the body into major functional areas. The organ represents only a particular region in that area, the shape the tissue assumes in one sector of the system. Whatever the position of an organ and its relation to the surrounding area, it has to be related both to its own system, in order to determine its role, and to its own tissue, in order to understand its properties.

However, when the texture, thickness and activity of tissues rather than their outward appearance are examined, the apparent variety of tissues becomes reduced to a small number of types – between ten and twenty-one, according to anatomists. For Bichat, Nature is always found 'to be uniform in her procedures, variable only in her results, sparing in the means she uses and generous in the effects she obtains from them, modifying in a thousand ways the few general principles that, differently applied, rule our economy and produce its innumerable phenomena'.[77] By classifying membranes in terms of their structure, properties and role, two main groups can be distinguished: simple membranes, 'whose separate existence is linked to neighbouring parts only by indirect organizational relationships'; and compound membranes that result 'from the assembly of two or three of the preceding and which unite their frequently very different characters'.[78] Ultimately, vital properties are located in tissues, and the attributes of the organism in the association between these tissues. Each element, each structure is cut out of tissue like a garment from a piece of cloth. Just as a living body consists of a collection of organs, each of which contributes to the properties of the whole by performing a function, so an organ is composed of several interwoven tissues, each of which, playing its own role, gives a large range of properties to the structure as a whole. A small number of tissues is therefore enough to provide variety in the living world. 'Chemistry has simple bodies which, by the various combinations they assume, form compound bodies . . .' said Bichat. 'In the same way, anatomy has its simple tissues which by their combinations . . . form organs.'[79]

With Bichat, therefore, a new level of organization emerged, intermediate between the organ and the molecule. The tissue represented the ultimate point of anatomical analysis, that to which a living body could be reduced by means of scalpel and scissors. The properties of an organism or its parts were not inherent in the molecules of the matter forming it. In fact, they disappeared as soon as the molecules were dispersed. It was the way these molecules were arranged in the form of tissue that gave living beings their intrinsic properties. The tissues were raw materials, each kind destined to perform a particular function. There were different tissues for cartilages and glands just as there were poplins for shirts and broadcloths for overcoats. The very word 'tissue' indicated the continuity of structure. What composed the living being were the uninterrupted layers and sheets that curled into organs, entwined, separated and wrapped together, extending from one structure to the next. The complexity of functional systems was resolved into the simplicity of anatomical combinations.

The continuity of the tissue corresponded in a way to the totality of the living being, on which biology put such emphasis at the beginning of the nineteenth century. A living body could not be divided into an infinite number of parts. It could no longer be imagined as a mere association of elements as Maupertuis and Buffon envisaged. Even when Oken again brought up the idea that beings were composed of elements, he did not contemplate autonomous units bracketed together, but units amalgamated in the wholeness of the complete organism. Oken's new idea, from which the cell theory was gradually to emerge, was to consider the bodies of large animals in relation to microscopic beings and to visualize the latter as elements of the former – in short, to imagine the complex living organism as an association of simple living organisms. Left to death and destruction, the flesh of animals and the tissues of plants decomposed into an infinite number of 'infusoria'. Each of these minute beings seemed to consist of a drop of mucus, the same sticky substance that was found in all living beings. For Oken, the small creatures thus released after death really represented the elements of which living beings were constituted: arranged in alveoli or cells, they formed the tissues of living bodies. However, for the animal to remain a whole, the cells were not merely heaped together like grains in a pile of sand. 'Just as oxygen and hydrogen disappear into water,' wrote Oken, 'and mercury and sulphur into cinnabar, a true interpenetration, an intertwining and unification of all the animalcules is produced here.'[80] There was no incompatibility, therefore, between the concept of an elementary composition of living beings and that of their totality, as long as an organism was considered as an integration of units and its decomposition after death as a disintegration. The elementary units could not merely associate and preserve their individuality in a complete being. They had to blend into a new individuality that transcended them. The parts were dissolved into the whole.

New ways of considering the fine structure of living beings were thus opened by the outlook of biology. In the eighteenth century, the idea of living particles united into an organism by their affinity had represented only one aspect of the combinative system underlying the formation of all bodies in the universe. Organic molecules had only been one particular category of molecules, reserved to living beings. Indestructible, each of them, once released by death, could always enter into a new combination and reappear in a living body. At that period the properties of an organism had represented the mere sum of the properties of each constituent molecule. In the nineteenth century, there was a total change. Interested in the organization of life rather than the form of living beings, biology was able to establish a relationship between the most complex and the simplest organisms. By considering the small organism as an element of the large one, it looked for the common denominator of all living beings, the basic unit of the animal or the plant, as it were. That unit could no longer be a simple molecule, an inert element or a portion of matter. It was itself a living body, a complex formation, able to move, feed and reproduce, a body, in fact, endowed with the principal attributes of life. But an animal or plant could not be a swarming mass of little independent beings. To consider an organism, with its unity, coordination and regulation, as composed of living elements, it had to be admitted that these elements were not merely stuck together, but integrated. The units had to be amalgamated in another unit of higher order. They had to be subject to the authority of the organism and renounce their individuality in favour of that of the whole organism. Only at this price could the indivisible being be composed of elementary units. The organism was not a collectivity, but a monolith.

Only after the possibility of such relations between a living organism and its component parts had been accepted could sense be made of the cells, the alveoli and the honeycomb pattern that had been seen in certain tissues ever since the seventeenth century. Within a few years, a whole series of observations were made both on the fine structure of plants and animals and on their reproduction. For, as Maupertuis and Buffon had shown, the study of reproduction could not be dissociated from the study of the organism's constitution. The importance of the cell theory lay in the fact that it provided a common solution for two apparently distinct problems: by resolving living beings into cells, each of which had all the properties of life, the theory provided both a meaning and a mechanism for their reproduction.

It was precisely at that time that the resolving power of the microscope began to be improved. The examination of tissues everywhere revealed vesicles, utricles, cells, more or less compacted, more or less cemented together, sometimes separated by channels. First in plants, where they could be more easily seen because of their size and shape; subsequently in animal tissues. Even certain microscopic organisms looked like cells. In an amoeba, for instance, there were no distinct organs, but according to Dujardin, a single drop of 'a glutinous, diaphanous substance insoluble in water, clinging to the dissecting needles, contracting in globular masses and allowing itself to be drawn into threads like mucus'.[81] That substance, first named 'sarcode' by Dujardin, was later to be called 'protoplasm' by Purkinje and von Mohl. When heat and acids were used to dissociate plant tissues, the same vesicles, the same globules enclosed in membranes always appeared. Various solutions were used to impregnate tissues in order to stain preferentially certain areas of the cell. Gradually the cell emerged, not as a mere drop of mucus, but as a small structure in which different areas, cavities and granules in constant motion could be seen. Above all, each cell was observed to contain a darker and denser single mass, which Brown called the 'nucleus'. Irrespective of form and function or of position in the organism, in both plants and animals, a cell always seemed to have the same general appearance, as if it always conformed to a single plan.

For the eye armed with a microscope, every living organism was finally resolved into a collection of juxtaposed units. This was the conclusion reached by most histologists, and which was generalized for plants by Schleiden and for animals by Schwann, under the name 'cell theory'. The cell theory, however, was not solely concerned with the problem of structure. For Schwann, the cell was no longer merely the ultimate limit in the analysis of living beings. It became both the unit of life – the constituent that possesses all living properties – and the point of departure for every organism.

The elementary parts of all tissues are formed of cells in an analogous, though very diversified manner, so that it may be asserted that there is one universal principle of development for the elementary parts of organisms, however different, and that this principle is the formation of cells.[82]

What was important was no longer the fact that cells were found in all tissues, or even that all organisms consisted of cells; it was that the cell itself possessed all the attributes of life and that it represented the necessary source of every organized body.

For this very reason, the cell theory struck a blow to the vitalism that had been prominent during the foundation of biology, by rejecting one of its fundamental postulates. In order to distinguish the living from the inanimate, it had been necessary to consider each being as an indivisible totality. The zoologists, the anatomists and the chemists, all had maintained that life resides in the organism as a whole, and not in this or that organ, part or molecule. Irreducible to simple elements, life remained inaccessible to analysis and interpretation. Hence the requirement for a continuity in the intimate structure of organisms, a continuity which Bichat found in the texture of tissues, and Oken in the fusion of cells into an 'infusorial mass' where the individuality of each element was submerged. It was precisely such ideas of totality and continuity that Schwann contested when he considered, not simply the elementary structure of living organisms, but the causes that determine two of their principal properties: nutrition and growth. For the vitalist, the causes of these two phenomena reside in the whole organism. It is the combination of molecules in a whole that generates a force allowing the organism to draw on surrounding materials and obtain from them the constituents necessary for the growth of all its parts. No part taken separately, therefore, is able to feed itself and grow. But it can equally well be maintained that in each cell the molecules are so arranged that the cell can attract other molecules and grow by itself. The properties of life can then no longer be attributed to the whole, but to each part – each cell – which in some way possesses an 'independent life'.

Schwann contended that all the observations made on plants and animals confirm this second interpretation. What is the egg of an animal, if not a cell capable of expanding and multiplying of its own accord – particularly the eggs of females that reproduce by parthenogenesis, since in their case it is impossible to evoke a mysterious force supplied by fertilization? What is a spore that gives rise to certain lower plants? And, in some plants, can fragments not be removed that are still able to multiply separated from the parent organism? There is therefore, no reason to endow the whole plant with special properties. 'The cause of nutrition and growth resides not in the organism as a whole, but in the separate elementary parts – the cells,' concluded Schwann.[83]

The resolution of the organism into elementary units does not accordingly destroy its ability to feed, grow and multiply. The specificity of life is not the attribute of the organism as a whole. 'Each cell leads a double life,' wrote Schleiden: 'one independent, pertaining to its development; the other intermediary, since it has become an integrated part of a plant.'[84] The organism can no longer be considered as a monolithic structure, a sort of autocracy whose powers are not shared by the individuals it governs. It becomes a 'cellular state', a collective system in which, Schwann said, 'each cell is a citizen'. Although constructed on the same plan, the different cells of an organism are of different types and perform different functions in different tissues; each type carries out a particular mission for the whole community. In the cellular community, there is a distribution of tasks and a division of labour. The existence of an organism, therefore, depends on the cooperation of its parts. Although the organism determines the conditions, it is not the cause of its own existence.

The properties of life, therefore, must perforce reside in the cell, not necessarily because of some mysterious force at the service of a divine intelligence, but rather through a particular arrangement of the molecules that allows the cell to effect certain chemical reactions. According to Schwann,

These phenomena may be arranged in two natural groups: first, those which relate to the combination of the molecules to form a cell and which may be denominated the *plastic* phenomena of the cells; secondly those which result from chemical changes either in the component particles of the cell itself, or in the surrounding cytoplasm, and which may be called *metabolic* phenomena.[85]

The cell can be considered as an individual separated by its membrane from the rest of the world. Nevertheless, although the membrane isolates the cell, it also allows the cell to enter into relationship with its environment, take nourishment from it and excrete waste-matter. The membrane has to be endowed with special qualities, in order to account for its ability to distinguish between the contents of the cell and the surrounding environment. One must ascribe to the membrane, said Schwann,

not only the power in general of chemically altering the substances which it is either in contact with, or has imbibed, but also of so separating them that certain substances appear on its inner and others on its outer surface. The secretion of substances already present in the blood, as, for instance, of urea, by the cells with which the urinary tubes are lined, cannot be explained without such a faculty of the cells.[86]

For Schwann, there is no need to invoke a mysterious force. An electric current is known to cause the decomposition of certain substances and separate their constituents. Why should the properties of the membranes not be due to the position of the constituent atoms? To explain vital phenomena, it is sufficient to postulate forces that operate, like physical powers, 'in accordance with strict laws of blind necessity'.[87]

The second aspect of the cell theory was concerned with the production of cells and organisms. This was no longer a question of composition and structure, but of origin. Everyone could see a cell divide into two, just as Siebold saw the division and multiplication of Protista, those organisms composed of a simple cell. The organism is, therefore, comparable to a colony of Protista and it is through a series of cellular divisions that an animal's body develops from the egg. Schleiden and Schwann assumed, however, that although a cell population multiplied by cell-division, each cell was not necessarily derived from another cell. Under certain conditions, cells might also be formed by a sort of spontaneous generation, starting from a 'primitive blastema'. Only with the work of Virchow was the possibility of a cell arising from organic magma to vanish. 'Where a cell exists there must have been a pre-existing cell, just as the animal arises only from an animal and the plant only from a plant.'[88] The continuity of forms and properties through successive generations applied not only to animals and plants, but also to the units that constitute them. The cell theory then assumed its final form as summed up in Virchow's words: 'Every animal appears as a sum of vital units, each of which bears in itself the complete characteristics of life.'[89]

With the cell theory, the composition of beings and their properties were no longer based on the demands of a system, but on objects demonstrable by observation. In the end, analysis succeeded in confirming the logical necessity of a combinative system, such as Maupertuis and Buffon had searched for. Whatever the nature of an organism, whether an animal, a plant or a micro-organism, it is always made from the same elementary units. It is the arrangement, number and properties of the cells that give the organism its shape and qualities. Even though the cell is a highly complex object, the same principle of construction operates, both in the living and in the inanimate world. With the cell, biology discovered its atom. Every aspect of the study of living organisms was transformed by the cell theory. To characterize life, it was thenceforth essential to study the cell and analyse its structure: to single out the common denominators, necessary for the life of every cell or, alternatively, to identify differences, associated with the performance of special functions.

The aspect of biology most profoundly transformed by the cell theory was the study of reproduction. Until that time, nineteenth-century biologists had simply continued the investigations begun during the preceding century. Gradually, improvement of the microscope and increasingly rigorous observation brought to a close the old argument about preformation and epigenesis. By filtering male spermatic fluid, Prévost and the chemist J.-B. Dumas had shown definitively that animalcules – 'zoosperms' – were necessary for fertilization. It had been recognized that the female egg was not the follicle which de Graaf had seen in the ovary, but a whitish mass detected inside the follicle by von Baer. The development of the embryo after fertilization became an object of systematic study, and the phenomena already described by C. F. Wolff close on a century before were again demonstrated, in greater detail. What von Baer saw after fertilization of the egg was not the development of a little preformed organism, but a series of complex events from which the shapes and structures of the future adult gradually emerged. The egg is only a sort of small ball at first which 'segments' into two, then four, then into a large number of alveoli stuck together. Progressively folds are formed, 'leaflets' that slide over each other, roll up, buckle and become indented to form the organs. 'Vertebrate development consists in the formation, in the median plane, of four leaflets,' said von Baer,

two of which are above the axis and two below. During this evolution the germ subdivides in layers, and this has the effect of dividing the primordial tubes into secondary masses. The latter, included in the other masses, are the fundamental organs with the faculty of forming all the other organs.[90]

In a given species, the development of the embryo always occurs in the same way, in the same order in time and in space, as if following a certain plan. There is first a progressive definition of the animal body, as differentiation of its form and tissues increases. Then the shapes that have been roughly sketched in are gradually refined, assuming the form of more specialized structures.

When neighbouring species within the same family were compared, remarkable similarities in development were found. That was what von Baer called the law of embryonic resemblances.

In my possession are two little embryos in spirit, whose names I have omitted to attach, and at present I am quite unable to say to what class they belong. They may be lizards or small birds, or very young mammalia, so complete is the similarity in the mode of formation of the head and trunk of these animals. The extremities, however, are still absent in these embryos. But even if they had existed in the earliest stage of their development we should learn nothing, for the feet of lizards and mammals, the wings and feet of birds, no less than the hands and feet of man, all arise from the same fundamental form.[91]

The same applies to the grub-like larvae of butterflies, flies and coleoptera: the larvae are often more like each other than are the perfect insects.

The discontinuity in the living world revealed by comparative anatomists was also found by embryologists through comparative studies of development. When comparing the segmentation of the egg, the movement of the layers and the order in which the different organs appeared, von Baer did not find a continuous series throughout the animal kingdom, but rather groups or 'types' of development. In fact, he observed four principal types, corresponding to Cuvier's four branches. From one group to another, embryonic development was radically different, while within a single group, it was similar. There again, the constraints of organization limited the number of possible variations, not only in space, but also in time. It was the most important, most central and deepest operations that occurred first. 'The most general features of a large group appear in the embryo before the more special features,' wrote von Baer. 'The least general structures are born of the more general, and so on until in the end the most special features appear.' The vertebrate could be recognized in the embryo before the bird. There seemed to be an embryonic history: to form basic organization shared within a given group, it seemed that all members of the group started out on a common path, the less perfect members stopping en route, before the more perfect ones.

The egg could therefore no longer be regarded as a rigid structure from which an already prepared form came forth. It was the origin of a system in which a whole series of successive alterations was taking place, each stage carrying the possibility of the next one. During its embryonic development, the living organism appeared to be formed of a succession of events, each generated from a preceding one, as if the organization unfolded in time as well as in space. Not merely the time of individual development, measured by the gliding of layers and the appearance of organs; but also a time that was more distant, more obscure, more profound and which seemed to portend a new kind of relationship between certain living beings.

The study of developmental anomalies then acquired a new importance. For, with Broussais, a new means of investigating living beings had appeared. Experimentation in physiology had usually involved modifying the natural state of an organism in order to disturb one function or another. However, the same result can be obtained by observing certain pathological conditions. What is disease, save an exaggeration or deficiency of certain processes which occur in a healthy animal? In many cases, no experiment can reproduce a departure from normality so precisely and selectively. If knowledge of the physiological state was obviously necessary for the interpretation of pathological conditions, the study of pathological conditions also provided a precious instrument to study biological functions.

Accordingly, the status of monsters also changed. Their formation could no longer be ascribed to divine anger, punishment of a secret fault or retaliation for an unnatural act or even thought. Nor could these beings on the fringe of natural order have been prepared from the beginning of time, awaiting with the others their turn to see light of day. For the new biology, it is during the course of embryonic development that deformities are produced, by some damage to the embryo. If hen's eggs are shaken violently during incubation, chicks afflicted with all kinds of anomalies are hatched. Until that time the monster, according to Étienne Geoffroy Saint-Hilaire, expressed 'the manifestation of Organization on days of saturnalia, weary of having industriously produced for too long a time and seeking amusement by abandoning itself to whims'.[92] Henceforth it becomes a 'being that has been injured during foetal life'. Monstrosity is simply the result of harm done to the embryo, failures in the succession of events by which an animal is formed, errors in the execution of the plan. Anomalies, irregularities and defects in formation result either from arrested or from retarded development. The idea of outlandish beings and products 'created without God', as Chateaubriand said, is supplanted by that of beings hindered in their development, whose organs remain in the embryonic state until birth. Even deformities themselves do not appear to occur at random. They affect certain areas of the body more readily than others, changing the structure, as if development had merely deviated from its course. There are rules in anomaly. 'Monstrousness is no longer a random disorder,' wrote Isidore Geoffroy Saint-Hilaire, 'but another order, equally regular and equally subject to laws: it is the mixture of an old and a new order, the simultaneous presence of two states that ordinarily succeed one another.'[93] All types of monstrosity are not possible. The deformities observed conform to certain types. Like normal beings, monsters can be classified. Anomalies obey certain rules of coordination, correlation and subordination. Some anomalies can be transmitted by heredity, but most result from damage to the embryo during foetal life. Teratology, the study of monsters, was to provide biology with one of its main tools of analysis.

With the study of embryonic development and the successive stages observed in the egg, a long step had been taken from the old theories still current at the beginning of the century. But it was not possible to interpret the shapes perceived in the egg, until they had been compared with those of cells in tissues. If the observations concerning the reproduction and development of the embryo contributed to the establishment of the cell theory, this theory made a reciprocal contribution, by providing explanations of many different aspects of generation. After all, like the egg, the 'zoosperm' was only a cell, although peculiar in shape and function. Fertilization was therefore the fusion of two cells, one from the father, the other from the mother. The segmentation and formation of germ layers as seen under the microscope resulted from cell division and the progressive differentiation of the cells which prepared to accomplish distinct functions and to form the organs. This led to the conclusion, Remak wrote,

that all the cells or their equivalents in the fully developed organism have arisen by a progressive segmentation of the egg cell into morphologically similar elements; and that the cells which form the early basis of any part or organ of the embryo, however small their number, provide the exclusive source of all the formed elements (i.e. cells) of which the developed organ consists.[94]

The formation of a living organism is thus indeed a re-production, an edifice reconstructed at each successive generation. But even if no longer attributable to the development of a little preformed creature, it is not total epigenesis, not a sudden organization of hitherto coarse material. 'Organized bodies,' said von Baer, 'are neither preformed nor, as is commonly supposed, do they come forth suddenly, at a particular moment, out of a formless mass.'[95] Every organism always originates in one of those units that form the living, a drop of protoplasm wrapped in its envelope, that is, a structure already possessing all the attributes of life. There are many ways of reproducing: by fission, as in Protista; by parthenogenesis, as in some aphids; or by the fusion of two germ cells, one paternal and one maternal, as in most animals and plants. But irrespective of the mode of reproduction, it is always from a fragment of organism that a new organism arises. Life is transmitted by a scrap of the parents that breaks away from them, develops and multiplies on its own, to reproduce an organization similar to the one from which it came and to acquire independence. There is never a complete rupture between one generation and the next; an element always persists, a cell that progressively blossoms into an organism. 'Throughout the whole series of living forms, whether whole animal or plant organisms, or their component parts, there is an eternal law of continuous development,' wrote Virchow.[96] Life is born of life, and of life alone. The formation of an organism by another is always a proliferation of cells, a sort of budding. A child is never more than an outgrowth of its parents. It is the cell that ensures the continuity of life.

Every organism, therefore, originates from a unit taken from the preceding generation. That unit divides by segmentation, and the cells thus formed differentiate to perform different functions, combining in tissues and organs, shaping the structures from which the architecture of the animal or plant gradually emerges. Each organism is a clone, an association of cells of various types, but all descending from the original cell, the egg. It is the number of cells, the variety of their types and their arrangement that determine the form and properties of an organism. It is by the combination of cells that the diversity of the living world is ensured. However, when a cell separates from one organism to form another, the new organism always develops in the image of the old one. From one generation to the next, all the processes of development, cell division and differentiation are unfailingly repeated, so that the same structure and the same system always emerge from the egg cell. The plan of organization revealed by comparative anatomy has necessarily to rest on a plan of development that directs the multiplication, differentiation and arrangement of cells in the egg. If the plan is consistent, if it is executed according to the rules, the child is normally formed in the image of its parents. If the plan is defective or poorly carried out, deformities appear. Biology thus encountered the same problem already perceived by Maupertuis and Buffon: the reproduction of an organization by assembling elementary units requires a 'memory' transmitted from one generation to another. In the first half of the nineteenth century, only the 'vital movement' could play the role of memory and ensure faithful reproduction. But whatever the name and the nature of the forces which transmitted the organization of the parents to the child, it was in the cell that they had henceforth to be found.

*

With the substitution of organization for visible structure as the object of analysis, the study of living systems acquired a frame of reference into which the data of sensory perception could be inserted. In the nineteenth century, organization was identified with life because it constituted a meeting-point for three interdependent variables: structure, function and what Auguste Comte called '*milieu*' [environment]. Life could exist only in so far as these three parameters remained in harmony. Any variation in one of them influenced the whole organism, which reacted by modifying the others. In a defined environment, wrote Comte, 'given the organ, find its function, and vice versa'.[97] This interaction was henceforth to provide the basis for analysis of the functions and properties of living systems. The study of biology consisted in examining the variations produced in certain parameters in response to changes, whether natural or provoked, in another. From that time onwards, all the efforts, ingenuity and procedure of biologists aimed at finding the means of isolating one variable, inventing a method for altering it in a predetermined way, and measuring the effects on the others. Henceforth, there were only 'three-body problems' in biology.

With the articulation of structure, function and environment, the manner in which organisms were arranged in space was entirely changed. The first change concerned the space occupied by the living world as a whole: in organizational terms, all combinations of elements were no longer possible. Only those arrangements that satisfied the conditions of existence could live. Only those that were adapted to the environment could reproduce. The uninterrupted chain of organisms from end to end of the living world was replaced by a few major types of organization, a few masses isolated from one another. The continuity of life was no longer horizontal, but vertical: the succession of generations linked through reproduction.

Secondly, there was a change in the space occupied by the individual organism. In the small number of permissible combinations, the organs were not combined at random but disposed according to a precise plan. The essential organs were deeply buried in the inmost recesses of the organism, the minor ones displayed on the surface. What was important, what lay at the very basis of life itself and could not be changed without dramatic consequences for the animal's existence, was thus removed from the environment, and sheltered from all external influences. What was secondary, however, was in direct contact with the environment, and subject to its action. Since the secondary could vary, if not with complete freedom, at least to a certain extent, it was the site of all interactions between the organism and its environment. It was on the surface, in that envelope which both separated and united the organism and its environment, that external influences could have a more or less direct and lasting effect on the living being.

There was also a change in the space occupied by the actual substance of living beings, since they were always composed of cells. The continuity of fibre or tissue was replaced by the discrete nature of cellular arrangements. Instead of solidly woven textures and closely superimposed layers, living bodies became combinations of elements and clusters of units. With the advent of the cell theory, biology was given a new foundation, since the unity of the living world was no longer based on the essence of beings, but on their common materials, composition and reproduction. The particular quality of the living reflected its intrinsic organization. At the same time, however, the cell theory drew the living world closer to the inanimate world, since both were constructed on the same principle: diversity and complexity built up by combinations of simple components. The cell became a 'centre of growth', in the same way as the atom represented a 'centre of force'.

The final change was in the space linking successive generations, since the organization no longer sprang forth ready-made from seed, but gradually developed from a single cell separated from the bodies of its parents. During embryonic development, the growth of the egg generated a series of different organizations, of which only the last corresponded to the adult. Reproduction was no longer based on the persistence of immutable structures, but on cycles of successive organizations that linked the egg to the hen just as much as the hen to the egg.

Into space, so modified, time had also to be inserted. For when certain elements had to be organized and not merely associated, the problem of their genesis could no longer be stated in the old terms. Organizations were not linked to one another by their proximity in space nor by the analogy of their elements. They were linked by the succession in time by which relationships were set up between these elements. If two organized systems showed some analogy, it was because both had gone through a common stage in the series of successions. The idea of organization was indissolubly linked to that of its history. But the history of an organized system was not merely the series of events in which the system had been involved. It became the series of transformations by which the system was progressively formed. Thus the permanence of primary structure no longer had to be ensured from one generation to the next, since the same organizations could derive from one another by the same sequence of successive changes, that is, by the same 'evolution'. There was, in fact, a correlation between space and time during embryonic development. Organs were not formed in haphazard order, any more than they were distributed in the individual in haphazard manner. The most important organs were the ones most deeply implanted in the organization, the ones least easily modified and the first ones to be formed. Accessory organs, in contrast, were the ones at the surface, the most variable ones and the last ones to be formed. A plan of formation in time, therefore, corresponded to the plan of organization in space. The series of transformations that constituted ontogenesis thus indicated a new relationship between species of the same group. For in all the embryos of the group the important and hidden organs appeared first, as though to form the same basic organization. Only later did the embryos of the group diverge, as if the different species which had set out along the same road had stopped at different points, the most perfect going further along the road to complete the surface details. Behind the time of ontogenesis, another, more distant and powerful time was dimly perceived, hinting at a network of relationships between living beings. A theory of evolution then became possible.

3
Time

Time is much more than a mere parameter of physics for the biologist today. It cannot be dissociated from the genesis and evolution of the living world. Every organism encountered on this earth, even the simplest, is the product of an uninterrupted series of organisms extending over the last two thousand million years or more. Every animal, every plant, every microbe is merely a link in a chain of changing forms. Every living being is inevitably the result of a history that includes not only the sequence of events in which its ancestors participated, but also the succession of transformations by which that organism has been gradually fashioned. The idea of time is inseparably linked with the notions of origin, continuity, instability and chance. Linked with origin, because the appearance of life is considered as an event that occurred perhaps only once, and certainly very rarely: consequently, all existing beings descend from a single, common ancestor, or from a very small number of primitive forms. Linked with continuity, because since the appearance of the first organism, every living being is necessarily derived from another living being: only as a result of successive reproductions is the earth today inhabited by a diversity of organisms. Linked with instability, because although reproduction is an accurate mechanism, although it almost always leads to the formation of an identical organism, yet it inexorably produces from time to time something different: this narrow margin of flexibility is sufficient to ensure the variations needed for evolution. Linked with chance, finally, because no intention of any kind can be discerned in nature, no concerted action by environment on heredity that might direct variation into predetermined paths: consequently, there is no *a priori* necessity for the existence of the contemporary living world. Every organism, whatever it may be, is thus indissolubly joined, not only to the space which surrounds it, but also to the time that has moulded its present structure and that gives it a sort of fourth dimension.

For contemporary science, the present state of the living world is explained by evolution. The role that is attributed to the past, however, is necessarily conditioned by the way in which the present is viewed and interpreted. The weight and function that any given epoch attributes to time is determined, in fact, by its picture of the nature of things and beings, of the inferred relations between them. It is no exaggeration to say that until the eighteenth century living organisms had no history. The generation of a living being was always an act of creation, whether an isolated act requiring the immediate intervention of some divine power or a primal act carried out in series concurrently for all organisms destined to ever appear. Even when the species became more strictly defined, it was viewed as a kind of fixed framework filled by a succession of individuals. In the course of generations, the same shapes were always found in the same places. The picture remained frozen and unchanging. What could be the history of a preformed being, waiting to see the day in the loins of its successive ancestors?

Cataclysms

Only during the eighteenth century did the notion of time introduce itself into the living world. First, because the idea of reproduction gave living beings a past: the line leading back through parents, grandparents and all the monotonous ancestry by which the species was perpetuated. Secondly, and principally, because through the writings of Burnet, Woodward, Benoît de Maillet and above all Buffon, a series of cataclysms that had thrown this world into confusion became evident. The earth had not been constant since creation: suddenly it acquired a history; it had an age marked by epochs. In the eighteenth century, the succession of generations still seemed no more than a tedious chain of identical productions, a uniform line without ups and downs. The history of the earth, in contrast, appeared to be a series of catastrophes, a cascade of transformations spread out over long periods. It no longer concorded with the biblical account according to which, ever since Genesis, peace on the earth had been disturbed only by the Flood. The chronology of generations marked the time proper to living beings: it constituted their intrinsic time-scale, as it were. On the contrary, the time imposed on the living world by the cataclysms of the earth remained extrinsic to the living beings themselves. Organisms were only affected secondarily, to the extent that the vicissitudes of the earth disturbed their living quarters, climate and food. The primary effect of the cataclysms was that the surface of the earth, originally incandescent, gradually cooled down; covered with a sort of universal ocean, the earth's crust folded and bulged; continents emerged and mountains reared up; and, as certain lands subsided, new seas surged forth. So climates changed: at first uniformly warm, they became cold in some regions and temperate in others. All these events had repercussions on the living world by modifying, not the organisms themselves, but their distribution on the surface of the globe. The traces left by fossils testified to their identity with existing organisms. 'In inland districts,' said Buffon,

on mountain peaks and in places farthest from the sea, shells, skeletons of sea-fish and marine plants are found, which are just the same as the shells, fish and plants now living in the sea, and which are, indeed, exactly the same . . . There can be no doubt as to their perfect resemblance and the identity of the species.[1]

Dispersed by the cooling of the earth, species which had once flourished in warm climates were obliged to flee and assemble in the only region that remains warm today. The area that cooled down became gradually inhabited by organisms unknown elsewhere. And in the process, many species perished. In the long run, it is because of the ruggedness of the earth's crust, the variety of climates and the distribution of oceans and continents that the living forms now found on earth have been preserved. Without this geographical diversity, there would be no verdure, none of the lushness we see in the country and the forests. 'A gloomy sea would cover the whole globe,' said Buffon, 'and all that would remain of the earth and its attributes would be a dark abandoned planet destined, at best, to be the home of fishes.'[2]

Prior to this time, the immobility and rigidity of the living world had never been questioned. It could not even be imagined that the picture presented by the series of living forms could ever have been different from what it is today. But once it was realized that the earth had had a somewhat chequered history, a sort of tremor made itself felt under the living world. The pedestal on which that world reposed began to move. It suddenly appeared that, after all, organized beings might not necessarily be always immutable, that species could perhaps change in the process of time and organisms undergo transformation. The eighteenth century is often credited with the origin of transformist thought. The ideas of Linnaeus on the fixity of species are contrasted with an evolutionist trend that includes not only Buffon and Maupertuis, but also Benoît de Maillet, Robinet, Charles Bonnet and Diderot. It is assumed that this latter trend developed and expanded throughout the second half of the eighteenth century, remaining merely to be defined and exploited by the following century. It is essential, however, to make clear what is meant by the words 'transformism' and 'evolution'. A new outlook certainly became manifest in most mid eighteenth-century writings: there could be a passage from one living form to another; many species that had once lived had disappeared, often leaving only traces difficult to recognize and decipher; no one could assert that the existing animals and plants were stabilized for all time to come and would remain eternally the same. Without question, the past and future of living beings had become uncertain, a view expressed by Diderot:

The tiny grub wriggling in the mud is perhaps on the way to becoming a large animal; the huge animal that terrifies us by its size is perhaps on the way to becoming a grub, is perhaps only a particular and transient production of this planet.[3]

In the end, even Linnaeus became enthusiastic about the appearance of certain hitherto unknown flowering plants and foresaw the possibility of a 'transmutation' of species thanks to some kind of unnatural hybridization.

Nevertheless, the notion of transformation does not alone adequately define the concept of transformism. What characterizes transformism is an impulse derived from organisms themselves, leading them gradually from the simple to the complex through the

vicissitudes of the earth; it is the product of an ever-unstable equilibrium between living forms; it is a network of interactions between
organisms and their environment; it is the dialectic of similarity and
difference in a unified history of nature. Transformism, in short, is
a causal theory accounting for the appearance, variety and kinship of
species; and in this all-embracing sense, it cannot be found in the
writings of the eighteenth century. Biological time and terrestrial time
remained unconnected. Only exceptionally did they meet and inter-
act. All that one finds in the thought of the eighteenth century are
some of the scattered threads which were destined, in the following
century, to be united into the notion of a causal chain of biological
events.

Many writers raised the possibility of biological modifications
sufficiently profound to change one species into another. The
transformation of 'fish into birds', for example, was discussed at
length by Benoît de Maillet; and the importance which every evolutionary theory ascribes to the passage from marine to terrestrial
life is well known. Fish left behind the subsiding waters, said de
Maillet, were 'forced to get used to living on earth'.[4] The little fins
under the belly that fish used for swimming in the sea would have
developed into feet for walking on land. The beak and neck would
have grown longer. Bristles must have sprouted on the skin and
gradually become down and then feathers. Nothing was easier than
to imagine a winged fish, generally flying under water and sometimes
in the air, changing into a bird that always flew in the air but preserved
the shape, colour and gliding motion of the fish.

The seed of these same fish, carried into the marshes, might also have
given rise to this first transmutation of species, from ones living in the sea
to ones living on land. Even if a hundred million beings perished without
being able to acquire the habit, it was enough that two succeeded for
the species to arise.[5]

This sounds familiar enough to twentieth-century ears, conditioned
by more than a century of evolutionary thought. However, de
Maillet's principal concern was, given the presence of a universal
ocean in ancient times, to establish a correlation between the continuous series of organisms now living in water and those living on

land. Every land animal had an equivalent living in the water. There
was not only the correspondence between the bird and the flying
fish. 'The lion, the horse, the ox, the pig, the wolf, the camel, the cat,
the dog, the goat and the sheep all have their equivalents in the sea.'[6]
Each of these animals already existing under water could gain dry
land and take up its residence there at any time. That was the only
type of transformation discussed in de Maillet's book, the *Telliamed*.
There was no succession of variations, no linking in time, no increasing complexity and perfection of organisms as the earth grew
older.

On the other hand, writers such as Charles Bonnet or Robinet
inclined towards an increasing complexity in the living world, a
'progression' from a somewhat simplified state to another more
elaborate one. For Robinet, for instance, there was only 'one single
Being, the prototype of all Beings'.[7] There again, for us this expression might have a somewhat transformist air. However, the prototype
was, in fact, only a sort of living unit, of organic molecule used for
forming living beings. According to Robinet, a single plan of
organization or 'animality' is possible at the very beginning: it is the
plan that has been realized in the prototype. The latter has a natural
tendency to develop and combine to form gradually the most varied
kinds of beings. It is by combinations and variations of the prototype
that different living forms are produced and the continuous web of
beings is woven. Eventually all possible combinations are realized
and constitute all the links in the chain stretching from the simplest,
the prototype, to the most complex, man. Thus the chain is not
formed by the progressive thrust and the passage of one form into
another, but by a sort of combinative system that produces one
organism one day and another totally different one the next day. For
Charles Bonnet, on the contrary, all living beings wend their way
regularly towards a future state in which 'they will then be as different
from what they are now as the State of our Globe will differ from its
present State'.[8] Here again, there is no question of successive transformations leading through time from the simple to the complex, but
a general shifting of living beings, a translation of the whole living
world along the time-axis. While preserving its place in the chain,

each species can thus attain a new position according to its degree of perfectibility.

Then man, transported to another place more in keeping with the pre-eminence of his Faculties, will leave to the Ape or the Elephant the first rank he now occupies among the animals of our Planet . . . The lowest species, such as Oysters, Polyps and so on will be to the highest Species in this new Hierarchy what Birds and Quadrupeds are to Man in the present Hierarchy.[9]

Remembering that Charles Bonnet was a convinced preformationist, it is clear that this massive ascent of beings must have been planned since the Creation by the deliberate improvement of the germs nested in one another.

How arbitrary it is, therefore, to place such writings on the same level as those of Buffon or Maupertuis! Yet, even in Maupertuis's writings there were only a few elements of variation, never a complete theory proposing the progressive formation of species. It was Buffon who most clearly emphasized the importance of the conditions of life, climate and nutrition in determining how living beings fit them-selves into geological eras. For Buffon, organisms cannot be in-dependent of their surroundings. External factors act in two ways. First, by limiting the fertility of organisms. There is a certain stability in the living world, but it is the result of two opposing forces. 'The ordinary course of events in living nature,' said Buffon,

is generally ever constant, ever the same; its movement, ever regular, turns on two unshakeable points: one, the unlimited fertility given to all species; the other, the innumerable obstacles that decrease this fertility to a given degree, always leaving about the same number of individuals of each species.[10]

This smacks of Darwinism. It already contains the theme of limits imposed on the expansion of living beings, an idea to be taken up by Malthus and later still by Darwin and Wallace. With Buffon, how-ever, it is a matter of balance rather than competition. Equilibrium is thought of in terms of the harmony that must rule in nature, and not in terms of the struggles and drifts of populations. Although fertility is an inherent quality in living beings, it does not operate to vary populations, but to perpetuate types. The important thing is the maintenance of each species at a constant level. Second, in Buffon's view, external factors have also played another role in moulding the structure of the contemporary living world. If changes have occurred, they have come, not from organisms themselves, but from the effects the conditions of life exert on them, from the spasms of the earth and from the cataclysms that have eliminated certain living forms. External factors do not act directly on organisms to transform them. They do not act on populations to drive them along certain paths. They simply allow certain forms to live, and not others. The structure of the living world is not considered in terms of populations, but of types. Many species that have been modified 'by the great vicissitudes of land and waters, by the abandon or the culture of nature, by the long influence of a climate that had become adverse or favourable, are not the same as they were in former times'. On the contrary, in all the places where the temperature is the same, not only do there occur the same species of plants, insects and reptiles, 'without having been brought thither', but also the same species of fish, quadrupeds and birds 'without their having gone there'.[11] Since the same causes produce the same effects under similar conditions, organic molecules arrange themselves in a similar manner to produce similar forms. Yet, when organic molecules combine to form animals or plants, it is not to produce rudimentary beings, but, on the contrary, complex organisms already resembling those found today. All possible combinations are gradually realized. Most are successful and progressively fill all available spaces, every corner of the earth, every stretch of water: they gradually weave the continuous web of beings living on the globe at that time. Other combinations, in contrast, have miscarried. They have produced 'those defective monsters, those imperfect sketches planned a thousand times, realized by nature, and that scarcely having the power to exist could only have survived for a certain time'.[12] These faulty anomalies have therefore disappeared without leaving descendants, swept away like Diderot's monsters in 'the general purification of the universe'.[13] Although nature seeks to achieve every possible combination of forms and organs, certain combinations are not viable. 'Everything that can be, is,' said Buffon – but not everything can be. Fossils testify to the earth's past and monsters to the limits of nature. The great workman and nature's chief servant then becomes time. Yet time itself 'relates only to individuals'.[14] There can be variations that modify the species to some extent. There can be certain species such as the horse, the zebra and the ass, which obviously belong to the 'same family' with a 'main stem' from which 'collateral branches'[15] seem to radiate. There can even be a slight air of kinship between all the forms that breathe, a common 'basic organization', for instance, between man and horse. But, in the end, these variations always remain limited; certain types are as immutable as the universe, otherwise science would not be possible. That foundation of stability is ensured by the permanence of the interior moulds. 'The imprint of each species is a type whose principal features are engraved in indelible and permanent characters for ever more.'[16] Keeping in mind the changes that have occurred on the surface of the globe, putting the origin of animals at an epoch when the two continents had not yet been separated, admitting that in the new world certain animals had been transformed into new species – in short, by a little ready reckoning – the two hundred species of quadrupeds could be reduced to thirty-eight families created in the beginning.[17] Accordingly, at that period a compromise between creation and variation could be held responsible for the origin of the living world.

Buffon's attitude, however novel it may appear, was obviously far from being a real transformist theory. It did not suggest the formation of the complex from the simple. There was no progression of forms in time. When a species was transformed, it gained nothing. Variation meant a 'degeneration', a 'denaturation' by which organisms finally diverged from their original type and lost their purity of race. Even the effect of geography on beings still remained very far from what later became the interaction between the organism and its environment. There was as yet, in fact, no environment in the sense understood in the following century, no defined space surrounding the organism, extending it in some way, and acting on the organism as the organism acts on its environment. There were only regions of the earth that lent themselves better to certain forms of life; conditions of existence that could not support all types of organisms;

'circumstances' that modelled beings in somewhat the same way as they fashioned institutions for Montesquieu.

As for Maupertuis, he was primarily interested in the mechanics of variation. His writings first reveal the idea of internal changes transmitted by heredity bringing in their wake variations in the living world. These modifications occur in the interplay of elements ensuring the continuity of species by reproduction. The source of hereditary changes lies in the particles uniting at each generation to form the child in the image of its parents. Within that system, modification can be produced in two quite distinct ways. First, an excess or lack of particles for making a given part of the embryo might cause a change in that part. This is the explanation of superfluous fingers, or of albinos, those white negroes whose hair is like the whitest wool, whose eyes, too weak for the light of day, open only in the darkness of the night, and who 'are to men what bats and owls are to birds'. Once they have appeared, novelties of this kind often persist in families, sometimes skipping generations. The most dissimilar species

would have owed their first origins only to a few chance productions in which the elementary parts would not have maintained the same order they had in the parent animals; each degree of error would have made a new species; and by means of repeated divergences there would have been produced the infinite diversity of animals that we see today.[18]

But there is also another way of producing variations: by mating individuals of different varieties, for features of both parents appear in the offspring. Those who work to satisfy the taste of the curious are, so to speak, creators of new species. Races of dogs, pigeons and canaries then appear that have never been seen in nature. 'They are at first only chance individuals; art and continuous breeding have turned them into species.'[19] To satisfy the fashion of the moment, new species of plants and animals are invented every year, the shapes corrected and the colours varied. For Maupertuis, just as for Darwin a century later, what the breeder obtains by art is a model for grasping how nature produces new species spontaneously. In man, for instance, chance mating gives rise to sickly or handsome races, giants or cripples, beauties or cross-eyed offspring. The attraction inspired by some and the repugnance felt for others, however, determines whether they have offspring or not, and hence the preservation or disappearance of their exceptional features. The consumptive type is rapidly eliminated. Shapely legs, in contrast, often improve from one generation to another. Friedrich Wilhelm's taste for stalwart grenadiers succeeded in increasing the height of his subjects. By imitating breeders, it must, therefore, be possible to create new types of men. 'How is it that those sated sultans, who shut up only women of all known species in their harems, do not have new species made for them?'[20]

With these two mechanisms, the combinative system of visible forms in living beings corresponds to that of the particles used for reproduction in the seed. Anomalies in the supply of particles or faults in the formation of the embryo produce every possible kind of living being. Yet for Maupertuis, as for Buffon and Diderot, not every realization is viable. Among all the chance combinations formed by nature, the only ones that can exist are those with 'certain suitable relationships'.[21] Chance has given birth to a multitude of individuals. Only in a small number of these are the organs arranged in a way to satisfy the needs of the organism: they have survived. In most cases, on the contrary, some constitutional disorder prevents development: they have perished. 'Animals without mouths could not live, others lacking genital organs could not perpetuate themselves: the only ones that remained were those in which there was order and conformity.'[22] In this abundance of possible beings that result from infinitely diversified variations, it is nature that makes the choice in the long run.

Characteristically, Maupertuis carried to extremes a system in which visible structure was controlled by a structure of a higher order. He sought to base both the stability and the variability of beings on the idea that forms were reconstructed at each generation by assembling particles. These particles then had to be varied and arranged in an infinite number of combinations so as to produce all imaginable arrangements, types and varieties. Non-viable beings were formed just as much as viable ones, but only the latter were preserved after being sifted in nature's sieve. Even then the number of varieties perpetuated was sufficiently high to give its continuity to the living world. Time, however, played only a modest part in this affair. There was no progressive passage from the simple to the complex, no perfection by gradual stages, nothing that suggested a series of successive transformations by which the world then living on earth had gradually been fashioned. What interested Maupertuis was establishing a system intrinsic to living beings that, by the sole power of their constituent units, would be able to produce all imaginable varieties and run the whole gamut of possibilities.

One cannot, therefore, correctly speak of transformism during the second half of the eighteenth century. At that period it was important to harmonize the picture of a living world, hitherto considered immutable, with the revelation of the tremors that had shaken the earth. Despite the gaps revealed by fossils, despite the great migrations and cataclysms that had redistributed species like a pack of cards flung across the face of the globe, living beings still formed a continuous web: continuous because of similarities and differences in space, not because of affiliation in time. With the concept of reproduction, a new relationship appeared between beings, linking individuals vertically through generations. As yet, however, that relationship applied only to organisms belonging to the same species, or to species so closely related that their obvious resemblances required them to be grouped in 'families'. Although the idea of forms remaining rigidly fixed throughout generations was questioned, the flexibility attributed to living bodies affected only secondary components, only characters that were certainly important for the classification of organisms, but not for their general structures: height, length of ears or limbs, the number of fingers, the colour of the eyes or hair. In their main lines of structure, the species preserved their permanence. It could still not be imagined that the picture representing the living world had ever possessed a radically different form; it could still not be imagined that this picture had changed with time, apart from a touch here and there, some minor additions to cover up bare white patches on the canvas, slight modifications in the arrangement of the figures in space – in short, a few mere details. Even if the contemporary world could no longer be seen as absolutely identical with that established at the time of creation, its genesis still required a major act of creation. For the diversity of present-day forms to have developed, the principal types, the main themes on which nature had later carried out variations, must have been established at the beginning of the world. From the very start, at zero hour, there ought to be already enough species to form a continuous ladder. All that time could have done was to increase the number of rungs and bring them close together. There was no question of arranging the transformations undergone by living beings in chronological series. The rhythm given by succeeding generations and the landmarks testifying to the cataclysms that had shaken the earth did not yet combine to account for the gradual emergence of biological diversity and complexity. Although living beings begin to have histories, there was as yet no history of the living world.

Transformations

With the passage from the eighteenth to the nineteenth century and the advent of biology, it became possible to confer on time a function in the genesis of all living beings. First, because with the separation of the organic from the inorganic, all objects belonging to the class of the living were related. Secondly, because the continuity of the living, born only of the living, broke out of the rigid framework of species. Finally, because the relationships between organisms were no longer established in terms of their constituent parts or organs taken separately, but of their whole structure, by reference to a system of a higher order represented by organization. The degree of complexity of organisms and their level of perfection were then measured in what Lamarck called the 'principal masses' of the living world. Each mass had its own organization, its 'system of relationships' between structures which deteriorated step by step from the most complex to the most elementary organisms. It was the organs that varied, but not in parallel fashion and not in direct relation to the complexity of the organism. 'A given organ in a given species,' said Lamarck, 'reaches its highest degree of perfection; while another organ that is very poor or imperfect in this same species is perfected in another one.'[23] Any comparison of living beings, any classification that was not based on organization, was therefore necessarily erroneous and arbitrary. In contrast, by considering the masses, it was easy to recognize that continuous chain, that ladder of gradations running through the living world from the simplest to the most complex being. 'A unique and graduated series in the disposition of the masses exists for each type of living body, in conformity with the increasing degree of organization,' said Lamarck.[24]

Thus, it was organization as a whole that became subject to transformation. A system of relationships between the constituents of a living being was not necessarily immutable. It could be transformed into another system one degree more complex, by a one-way process. It then became possible to consider *all* living beings as deriving from one another, and to link them by a common movement through time, by a sort of impulse from within that tended to make living bodies more complex. By endowing organization with the capacity to transform itself, Lamarck was able to achieve what had been impossible in the eighteenth century – that is, to link all living beings by a single history describing their successive genesis. For Buffon, transformation applied only in very limited fields and occurred only within 'families' of species. At the origin of the living world, there had been about forty distinct types, from which new forms had sprung to make up the present living world. There was nothing, however, to unite these families with one another, no sort of affiliation and no kind of family tie. The word 'denaturation', used by Buffon to describe the variation of species, evoked a degradation, a corruption of the purity of the species. For Lamarck, on the contrary, transformation could only occur in the sense of adaptation, and thus of an increase in 'faculties'. Consequently, the different types of organization had not appeared simultaneously, but in a certain order in time, 'since in climbing the animal ladder, from the most imperfect animals, organization is composed and becomes more complex in a most remarkable manner'.[25] Time thus became one of the main operative factors in the living world. It was time that caused all forms to emerge gradually from one another. Beyond their diversity, all beings of the same kingdom became, therefore, linked by the unity of a common history. But that history could still only be represented by a single straight line without breaks or meanders.

According to Lamarck, three factors work together to give time its creative role: succession, duration and improvement of organization. In the first place, all the evidence indicates that all living forms cannot have been fashioned at the same time. Living bodies have experienced various degrees of change in the condition of their organs and the relations between them. The species, therefore, cannot be a rigid, unchanging framework into which individuals of successive generations fit themselves. 'What is called species . . . has only a relative constancy in state and cannot be as ancient as nature.'[26] This applies, not only to certain offshoots from a few original stems, but to all organisms. There is no longer any need for creation. 'All organized bodies are true products of nature that she has successively formed.'[27] Step by step all living forms have emerged by the same continuous movement in time, by a single series of consecutive transformations. There is a reciprocal relationship between one stage of transformation and the visible interval separating two adjoining levels of organization. If a spatial relationship can be established between the types of organization, then the time relationship of the transformations can be inferred: the former relationship results from the latter.

The series of transformations must have taken place over very long periods of time. All that is on the surface of the globe gradually changes its state and form. All bodies on earth undergo more or less rapid 'mutations'[28] according to their nature and the forces that act on them. The stability that man sees in nature is only on the surface, owing to the fact that he considers all events in terms of his own life-span. A few thousand years appear to him as an immensely long period; in fact, they only allow him to see stationary conditions, intervals between changes affecting the living world. Yet, even if modifications in living organisms are imperceptible to the eye, even if the forms seen in Egypt today are no different from those that lived there three thousand years ago, the slowness of the transformation process is compensated by its duration. For the diversity of the living world to have been born of such a slow process, it is necessary and sufficient to invoke enough time.

Compared with what we look upon as long periods in our everyday calculations, there must doubtless have been enormous time and considerable variations in circumstances for nature to lead the organization of animals to the degree of complexity and development that we see today.[29]

For according to Lamarck, transformation is a one-way process. Variation goes always in the same direction, from the simple to the complex, from the rudimentary to the detailed, from least perfect to most perfect. Any change that takes place in an organism and produces another one necessarily brings in its wake an increased organization, a greater aptitude for satisfying a need, a higher capacity of response to the requirements of life. Transformations lead only to happy issues, never to failures in the form of 'lost species'. Many of the fossils found resemble current forms. If there are any fossils that resemble no existing species, it is because man has destroyed them, or because they have been modified out of all recognition. But nothing disappears from the living world. The most ancient species exist side by side with the most recent. The simplest forms and the most complex derived from them are found together. Thus, there is a common basis for three of the parameters that can be observed in the living world: the time when an organism appeared, its degree of complexity and its stage of perfection. If one is known, then the other two can be inferred, since they represent three expressions of the order followed by nature in producing the animal and vegetable kingdom.

If it is true that all living bodies are products of nature, it cannot be denied that she could have produced them only in succession and not all together at one fleeting time; but if she formed them successively, there are grounds for thinking that she began with the most elementary forms and produced the most complex organizations last of all.[30]

Less perfected, therefore, also means earlier and less complex. This is the relationship permitting the series of organizations in space to be converted into an isomorphous series of transformations in time. To survey the continuous chain of beings, from the simplest to the most complex, corresponds to reproducing nature's progress in time, to reconstructing the succession of transformations by which organisms have been formed. On the ladder of living beings, the most elementary forms occupy a privileged position, since organization has begun by them. It is therefore among the most rudimentary organisms, the 'animals without backbones', that variations can be seen most clearly and the requirements of organization investigated most easily.[31]

At the time of its publication, Lamarck's *Philosophie zoologique* was coldly received. If Lamarck had some influence on his contemporaries, it was not so much because he proposed a genesis of the living world by consecutive transformations of one form from another, but rather because he detected, among living bodies, a unity reaching beyond their diversity; because he traced a boundary between the organic and the inorganic, and centred analysis on organization – in short, because he made a prime contribution to the establishment of biology as a science. Lamarck's writings, during the second period of his life, represented a transition between two forms of knowledge. Some of them already expressed the outlook of the nineteenth century, while others still reflected that of the eighteenth. For instance, all living beings still formed a continuous web, the old chain winding from one extremity of the living world to the other. Like Buffon, Lamarck did not believe that either species or genera really exist in nature. According to him, 'there are only individuals', forms 'that are close together, shading into one another and intermingling'. If a gap happens to occur in the gradation, it is because outside circumstances, anomalies in the way of life or changes in habits have disturbed the regular progress. Or it is because man's knowledge is incomplete and has not recognized all the links in the chain. The chasm that seems to separate birds from mammals is already closing, since the existence of intermediate animals such as Ornithorhynchus and Echidna has been recognized. In fact, it was because Lamarck still observed a linear series in the living world that he was able to interpret it as the result of a series of chronological events. It was because nature made no sudden leaps that relationships of similarity could be transformed into relationships of descent. 'She follows an order that is easily recognized, for it is exactly the contrary of what we observe when we look at all living beings from the most perfect to the most elementary.'

Again, with respect to the inevitability of the living world in its present state, Lamarck reflected the outlook of the eighteenth century. For him, certainly, the picture formed by living forms has not remained unchanged during the course of ages. The chain has not been complete ever since the origins of the world. Time has brought novelties. It has added links one by one, end to end. In the long run, however, the picture seen today could scarcely be different from what it actually is. Although he refuses to consider the living world as the result of an intention or the achievement of an aim by a supreme power, Lamarck attributes to animal life a 'prime and predominating cause' which enables it to make organization gradually more complex and more perfect. Due to this cause, the operations of nature develop according to a 'plan' which fits the formation of new beings into the condition of the world they have to live in. Before producing an organism, nature already knows what she has to produce. With vertebrates, for instance, it is obvious that 'nature had begun the execution of her plan with fish; then she took a step forward and made reptiles; next she further improved the plan and made birds, and finally completed it by making the most perfect mammals'.[32] Gradually, step by step, without errors or failures, nature creates, improves and modifies the forms of living bodies to achieve her end: the highest degree of perfection. Improving a system often requires a whole series of intermediates. To find the best solution for the problem of respiration, nature had first to perfect a system with a trachea, another with gills, before producing the most elaborate of all, the lung. There are no trials and errors, no hesitations or false starts. There are merely delays and complications whenever progress is 'thwarted' by external changes and circumstances.

Nature has at her disposal a combination of two factors for producing transformations: one is inherent to life, the other extrinsic. There is first, in each living being, a sort of force that 'ceaselessly tends to make organization more complex'. This is not the mechanism of reproduction itself, as Maupertuis would have had it, nor the irresistible impulse causing exponential multiplication of living beings, referred to by Benjamin Franklin and Malthus. It is a force of somewhat mysterious origin, a 'power' that, despite Lamarck's professions of materialism, is akin to the vital force: it is the prerogative of organized beings alone; it is the true source of harmony and regularity in the progress of living beings. But although this force perpetually increases organization, it is not alone sufficient to produce the diversity that exists among organisms. It is obvious, for instance, that if fish 'had always lived in the same climate, the same kind of water and at the same depth, then doubtless the organization of these animals would show a regular gradation'.[33] But then again, they might never have emerged from the water to populate the continents! For living forms to become diversified, conditions of life, 'circumstances' and 'environment' have to play a part; not in the way Maupertuis, Buffon and Diderot saw it – for them, external conditions only encouraged or impeded the perpetuation of already existing organisms – but by direct action on the properties, structure and heredity of living beings. For animals to leave the water, they had 'to be brought gradually from living in water to living at the water's edge, then on dry parts of the globe'. Once on dry land, as they experienced new desires, they developed new habits and acquired an organization better adapted to their situation. Cabanis had already proposed that needs or even desires provoke development of instruments for satisfying them – that is, organs. According to Lamarck, a whole network of interactions is established between 'the product of circumstances as the cause of new requirements, the effect of repeated actions creating habits and leanings, the result of increased or decreased use of any one organ, and the means used by nature to preserve and perfect all that had been acquired in organization'.[34] Although the inherent ability of all living beings to increase the complexity of their structures is sufficient to ensure the transformation and improvement of organisms, it is external circumstances that disturb their regular progress and start them on fresh paths.

The unending opposition between the faculties of the organism itself and external circumstances resulted from what Lamarck considered to be one of the most indisputable properties of living beings: their power of adaptation to conditions of life, the concordance between the organism and its surroundings. Lamarck's attitude, like that of the eighteenth century, was based on the necessary harmony of the universe. The living world was not only the best, but the only possible one. No crisis could ever arise between living beings and nature, no combat between organisms themselves for the conquest of territory, no struggle for existence of the kind already invoked by Malthus and later postulated by Darwin and Wallace. According to Lamarck, when a new being is produced that does not conform to the regular progression, it is an adaptation to special circumstances. Variations are always useful. But the power of the living to perfect organization is not always sufficient by itself to produce useful forms. Although heredity creates, it does not adapt. Its new productions and improvements are regular, without fantasy or deviation. They are unable to cope with the unexpected, with special circumstances. Hence the necessity for environment to act on heredity through desires, needs, habits and actions. Once the organization of certain individuals has been modified in this way, 'the generation between the individuals in question preserves the acquired modifications'.[35] The plasticity of living structures and the flexibility of their mechanisms allow the organism, not to insert itself into the surrounding world, but on the contrary to insert this world gradually into its heredity.

This was not the first time that the possibility of direct action by environment on heredity had been contemplated. From the earliest times, only the transmission to descendants of experience acquired by individuals appeared to account for the harmony between organisms and nature. Never had this idea been exploited so systematically and with so much detail, however – nor so confidently since Lamarck took for granted that an organ disappears because it is of no use. For him, whales and birds have no teeth because they do not need them. The mole lost the use of its eyes because it lives in the world of darkness. Acephalous molluscs have no head because they have no need for it. It is equally obvious that an organ develops because it is often used. The swan has a long neck because it feeds on acquatic animals. The duck has webbed feet because it swims by striking the water. Carnivorous animals have sharp claws because they have to climb trees, burrow into the earth and tear apart their prey. According to Lamarck, finality does not involve a primary intention, a decision to produce a living world and gradually guide its development. It is made of short-term finalities, so to speak, each centred on the well-being of an organism that is to be produced later, since adaptive intention always *precedes* realization. In the end, the plan followed by nature is aimed at providing the world with always more complex, more perfect and better adapted organisms. If the natural progression of organisms is not enough, it is corrected *in advance*. The execution of the plan results, therefore, in a series of small accumulated finalities. But although Lamarck describes profusely *why* things happen, he talks sparingly about *how* they happen. Mention is never made of investigations or observations designed to reveal a process that would permit external factors to act on heredity. There is no reference to the observations of Haller or Charles Bonnet that organisms remain constant, despite mutilations repeated in each generation. There is no suggestion of experiments similar to those already proposed by Maupertuis to determine whether heredity draws lessons from experience. There is only reference to 'internal fluids' acting on 'the flexible parts of the animal' to hollow out channels, displace masses or even construct organs; in short, gradually to shape the body. Lamarck's writings still contain an old flavour of mechanism, with the regularity and continuity of perpetual or uniformly accelerated motion.

In the last analysis, Lamarck's transformism has two motive forces. It is, indeed, 'evident that the state in which we find all animals is, on the one hand, the product of increasing complexity of the organization leading to regular gradation, and that, on the other hand, it is the result of the influence of a multitude of very different circumstances continually tending to destroy the regularity in gradation of the increasing complexity of organization'.[36] Hence there is first the continuous stream advancing along the highway –

the slow, regular and certain ascension leading large masses along nature's ladder. Secondly there are local movements, perturbations that, without affecting the mass, introduce 'deviations', 'divergencies' and 'digressions'. Finally, to keep the main current up to strength, there is a steady afflux of very simple organisms which are constantly formed from inorganic matter. At the bottom of the ladder, nature proceeds by 'spontaneous or direct generations that she renews ceaselessly whenever circumstances are favourable'.[37] She continually produces 'animalcules of the simplest organization' that 'one hardly suspects of having animality'. As soon as they are formed, these organisms join the main stream and begin climbing the ladder. The simplest beings are therefore also the most recent. At the other end of the hierarchy is man, who thus comes from the oldest organisms. Transformation goes no further. Yet, the living mass does not accumulate at the summit. Far from increasing indefinitely, the human population is subject to regulation. For as the flow of organic bodies ascends the ladder, those at the top return to the inorganic state. By degrees, the living regresses to the mineral. In that state, the material awaits the chance of some spontaneous generation to start climbing again from the foot of the ladder, as animal or vegetable. There might also be other points of entry for spontaneous generation higher up the ladder; at the level of 'intestinal worms', for instance, of 'certain vermin causing skin diseases', or of 'moulds, toadstools, even lichens'.[38] This kind of cycle keeps the living world in what we would today call a state of dynamic equilibrium. Nothing changes, for what disappears is exactly balanced by what appears. No rung of the ladder can be removed, no group of animals can die out. When a cataclysm creates a void, it is immediately filled up by the rising stream like a hole in water. As organisms move up and leave an empty space, others come to occupy it. The living world, consequently, cannot be modified. It keeps the equilibrium living beings impose on it. It always corresponds to pre-established harmony. Produced by a series of transformations, it has the immutability of a created world.

Lamarck, therefore, stands exactly at the point of transition from the eighteenth century to the nineteenth. Perhaps more than anyone else, he contributed to the change in outlook by which the living was separated from the inanimate and which established biology as a science. More than anyone else, he helped to make organization the centre of the living body, the place where all the component parts combine to let the whole function properly. Because he chose organization as the point where time acted, he was able, more clearly than his contemporaries Goethe or Erasmus Darwin, to envisage the whole living world as the result of successive transformations, as a progressive development of structures and functions. But however novel the description of a generalized transformism might appear, it was nevertheless founded on a representation of the living world still conceived in eighteenth-century terms. Lamarck's transformism was the linear chain of beings disposed in a linear sequence of time. The series of transformations was envisaged only through continuity of space. By this very fact, any fortuitous character in the arrangement of the living world was eliminated. Since they failed to find sufficiently strict arguments and critical observations in the *Philosophie zoologique*, Lamarck's contemporaries rejected his theory, even those who sought a common basis for the organization of all living beings. For the most part, they saw in it only a development of Buffon's ideas carried to an extreme, if not to the absurd.

Fossils

The aspect of Lamarck's transformism that belongs to nineteenth century thought is the motive force of time in the derivation of ever more complex organizations. It is the power of creation that the variety of organs taken separately confers on the system of their relationships. In the nineteenth century, in fact, the problem of genesis is posed in a new way: the component elements of objects are not merely juxtaposed, but linked by a whole network of relationships. These organizations, these structures of higher order that determine the arrangement of the various parts in beings and things and give meaning to both the parts and the whole can no longer have been set in place at the origin of the world according to a predetermined order. These organizations no longer precede a history that would have seized them at the moment of creation and handed them over to a time capable only of upsetting the initial order and hampering classification. The systems that can be observed have not been established once and for all; they have been formed at different levels and in successive stages by a sequence of events that have gradually articulated the components, arranging them in certain ways and constantly recasting their relationships to make them more complex. If there is a resemblance between certain types or organizations, it is no longer because of some pre-established harmony beyond man's knowledge; it is because they happen to have gone through some common stage in the series of transformations. Consequently, similarities in space are no longer enough to group the bodies in the world; their succession in time has also to be specified. What can be observed in any one domain cannot have been brought into being at a single stroke, or even foreseen in accordance with some preconceived plan. It develops by a continuous dialectical movement in which opposites become interpenetrated and quality is produced from quantity. The finality of the objects making up the world lies in their necessity, which itself can no longer be dissociated from their contingency. Instead of a chronological list of independent events, history becomes the movement in time by which the universe has become what it is, a process of development, a change from the most elementary to the most complex; in short, an 'evolution' born of the internal sequence of transformations. Many relationships between things or beings are thus reversed. It is evolution in time that determines relationships in space. Whether mind or nature is considered, their origin can no longer be regarded as a birth, a creation from which histories have developed under the influence of external impulses. Origin becomes the vanishing point of history, the necessary zone of convergence where all the rough outlines of organization meet, where all divergences, disparities and differences disappear.

The evolutionism of Darwin and Wallace is radically separated from all previous lines of thought by the notion of contingency applied to living beings. Up to the nineteenth century, the great chain of living beings was part of universal harmony. There was the

same necessity in the world of beings as in that of the stars. It was inconceivable that living beings could ever have been different from
what they were. But this necessity could not be brought into question
until the space occupied by the living world had been upset by
comparative anatomy, embryology and histology. There were
several reasons for this. First, a synthesis in time can be based only
on the image drawn from investigations in space. As long as the
living world had been arranged in the form of a continuous chain,
the succession of transformations could only be seen in Lamarck's
way, as a series of gradations moving in linear strides from the
simplest to the most complex. Once the continuity of the horizontal
relations between living bodies have been broken, once organisms
have been redistributed in 'branches', the vertical relation can no
longer be visualized as a single sequence. Then the isolated groups
of beings no longer bear the stamp of necessity.

Next, the cell theory wove a new network of both horizontal and
vertical bonds between organisms. The cell is at once the universal
component of all living bodies and the factor that unites one generation to the next. Every organism always develops by a sequence of
transformations in which cells multiply, differentiate and become
organized in a series of shapes. There are often analogies between
ontogeneses, sometimes even short stretches along a common path.
But there again, no necessity is involved, either in the path adopted
by development or in the way in which the cells are arranged and the
structure built.

The distribution of organs in a living body also led to an entirely
new way of considering the very possibility of variation. The
organism is curled round its centre in superposed layers extending
beyond the organism and inextricably linking it to its surroundings.
The further organs are from the centre, the less important they become and the more freedom they have to vary. The parts on the
surface become so secondary that there is practically no restraint to
limit their changes. According to Cuvier, for a modification to
appear, no 'form or condition is necessary; quite often it does not
even have to be useful to appear; it is sufficient for it to be possible,
that is to say, not to destroy the harmony of the whole'.[39] This is a

complete reversal of Lamarck's attitude. Variation in beings does not necessarily depend on usefulness, need or progress. It may be gratuitous.

Finally, the nature of relations uniting the organism to its surroundings was also modified. According to Lamarck, the 'surroundings' formed only one of the parameters of 'circumstances'. They defined the nature of the element in which an organism lies: air, fresh water or salt water. The animal was placed in certain surroundings, like any other body on the earth. But there was still no close relationship or real interaction between the organism and its surroundings. Although the surroundings often affected the structure of living beings, it was in a roundabout way, because of a need or a desire; because a being had to breathe, feed and move, whether in air or in water. With Auguste Comte, however, surroundings and circumstances became 'environment', and acquired a new status. Environment represented all the external variables to which the living being was subject – not merely air or water enveloping the organism, but also weight, pressure, movement, temperature, light and electricity – in short, anything that could affect a living body. This, however, was no longer a one-way process. The organism and the environment acted on each other. 'For according to the universal law of the necessary equivalence between action and reaction,' said Comte, 'the ambient system could not modify an organism without the organism exerting a corresponding influence.'[40] The organism cannot be dissociated from its environment. It is the whole system that is modified and transformed.

Consequently, there could no longer be one time-scale peculiar to living beings – that of their successive generations – and an external time-scale – that of the upheavals the earth's crust had undergone. In the eighteenth century, the time-scale of the earth, with its cataclysms, variations in temperature and disturbances of all kinds, had disrupted the order of beings in the monotony of their reproduction with no history. For Lamarck, in contrast, the time-scale peculiar to beings had created the progress of the living world, while the time-scale of circumstances only occasionally interfered to allow organisms to adapt and pattern themselves on their surroundings. In the nineteenth century, there could be only one time-scale for the universe as a whole. The history of beings was inextricably bound up with the history of the earth. Fossils then began to play a new role. In the eighteenth century, they had resembled existing forms and bore witness to the permanence of living bodies. For Lamarck, they helped to show the instability of living beings and indicated certain transformations. For Cuvier, they are milestones of geological time; the 'monuments of past revolutions' that have to be restored and deciphered. There is only one history, the history of nature, some-times told by stones, sometimes by fossils; and man has to learn how to collect and fit together all the marks. Fossils assure us that the globe has not always had the same covering as today, because we are certain that they lived on the surface before being buried in the depths of the earth. Rocks, in turn, reveal discontinuous layers in the earth's crust where vestiges of living beings were deposited. Only the application of the law of correlations to fossils and the reconstruction of a whole organized body from a fragment allow man to define the successive geological formations and to detect the cataclysms that generated them. Conversely, examination of geological strata describes the habitats of vanished species. If geology shows how certain continents are related, fossils tell how they were separated.

According to Cuvier, the thickness of different rocks and the grouping of different fossils from one layer to another constitute the traces of 'revolutions' that have suddenly overthrown the globe on several occasions. Animals could not have remained the same after these cataclysms. Breaks in the continuity of space occupied by living beings therefore merely reflect breaks in earth-time. 'Shells in the ancient strata have shapes of their own', said Cuvier,

and gradually disappear, being no longer found in recent strata, still less in the present strata in which analogous species are never found, nor even several of their own genera; shells in recent strata, on the contrary, resemble those of their own genera that live in our seas.[41]

There is, consequently nothing in common between what has lived in earliest times and what is living in the present. Cataclysms that have ploughed up the earth's crust have wiped out its inhabitants. Countless beings have disappeared in this way.

Those living on dry land were swallowed up by floods; others, peopling the deep waters, were left high and dry when the sea-bed was suddenly raised; even their races are ended for ever and all that remains in the world are a few fragments hardly recognizable to naturalists.[42]

There are no traces of a single line of descent for the living world nor of a series of changes by which each species might have gradually derived from another. There is no question of linking the simple to the complex by successive variations that would have progressively brought the animal kingdom to its present state. Nature has taken the greatest care to prevent the species being altered, keeping the main lines of organization fixed. Present races 'cannot be modifications of lost races'. Conditions formerly prevailing on earth cannot have been the same as those of today. And although Cuvier did not himself mention the notion directly, his disciples did not hesitate to speak, not of one creation, but of several, following each catastrophe in turn.

Cuvier's ideas about geological revolutions on the globe have exposed him to much criticism. They were – and are still today – often described as the expression of a retrograde, conservative and even theological attitude that retarded the advent of evolutionist thought. It appears quite clear, however, that the principal character of evolutionism, the contingent nature of the living organisms, could not be entertained as long as organisms were set in a rigid framework, progressing in Indian file towards perfection. The dispersal of living forms, the breaks in time that created them and the gratuity of variation were three preconditions for any theory of evolution. All three were the work of Cuvier.

It was the geologists who exorcised the devil of cataclysms. For dispersal into distinct regions and withdrawal into separate little islands were found, not only in the world of living beings, but also in the world of minerals. To account for geological formations, however, it was no longer necessary to invoke exceptional catastrophes. What Lyell called the 'principle of actual causes' was enough. All the evidence about its past borne by the earth shows that it has undergone a series of alterations. But these ancient changes produced in the earth's surface were brought about by factors similar in nature and intensity to those which operate today. According to Lyell,

To explain the observed phenomena, we may dispense with sudden, violent and general catastrophes, and regard the ancient and present fluctuations . . . as belonging to one continuous and uniform series of events.[43]

When the signs hidden in geological strata can be read, there is evidence that the conditions of climate, erosion and volcanic eruptions were the same in ancient times as at present. It is possible to see tree trunks upright with their roots still buried in the soil. In the mud and sand that hardened in olden days, ripples like those on our beaches can be discerned. Comparing, as Lyell did, the marks left by raindrops in rocks of all ages is like looking at imprints left in the sand by a number of showers a few days apart. The shapes and sizes of the drops are almost identical, showing that atmospheric conditions were similar. Neither rain, nor dust, nor deserts, nor ice, nor winds ever seem to have been different in the past from what they are today. Certainly, all evidence indicates that violent events have sculptured the earth's surface. Whole mountain ranges have surged forth from the breast of the earth or been buried in its depths; valleys have suddenly opened up, then been filled in, then opened up again; the seas have invaded the land, then withdrawn. But all these changes form a series of events without any real interruptions, a succession of epochs during which the earth's crust has gradually acquired its present configuration. By studying the arrangement of the masses of buried minerals, the geological order of the earth can be reconstructed. The rocks show an important chronological series of movements testifying to a succession of events in the early history of the globe and its inhabitants. Through all the transformations, and despite changes in conditions, circumstances and general or local climates, similar causes have led to the accumulation of the same material and formed similar strata. It has all happened without violating the laws which still control the formation of soils, sediments and rocks today. According to Lyell, geology 'consists in an earnest and patient enquiry of how far geological appearances are reconcilable with the effect of changes now in progress, or which may be in progress in regions inaccessible to us, and of which the reality is attested by volcanoes and subterranean movements'.[44] The past produced the present, but only the present can explain the past.

Rocks could no longer be classified just by examining their structure and texture. Their origin and age also had to be considered to deduce the order in which they were formed. There were three criteria for determining the age of a mineral mass: first, superposition, that is its place in relation to the surrounding rocks, the lowest being the most ancient; secondly, the character of the minerals, that are often identical in wide horizontal beds but change rapidly in vertical sections; thirdly the content of organic debris. According to Lyell, their very nature 'confers on fossils their highest value as chronological tests, giving to each of them, in the eyes of the geologist, that authority which belongs to contemporary medals in history'.[45] It is therefore in those formations which contain fossils that relative ages can most easily be determined. First a chronology of these formations is established, then as far as possible the different groups of rocks are related to these divisions. According to their origin and the causes of their production, rocks can be classified into four main groups: sedimentary, plutonic, volcanic and metamorphic. But these four classes of rocks were not produced in a single series. Each does not correspond to a specific epoch. This means that the succession of geological epochs cannot be reconstituted merely by examining the vertical order of the superimposed strata in a certain region. There is a sort of 'transmutation' of rocks, fossiliferous masses gradually being converted into crystalline masses. And this process, which is still going on today, has been in operation throughout time. Sedimentary and fossiliferous beds are being formed in certain lakes today, while volcanic formations are appearing elsewhere. In the same way, in each past epoch, fossiliferous deposits and igneous rocks were formed on the surface, whereas certain sedimentary strata, subjected to heat and pressure, acquired a crystalline structure. The relative position of rocks, therefore, can be compared only in limited zones because of the time-shift from one region to another. Consequently, the four main classes of rocks have to be considered, said Lyell, 'as four sets of monuments relating to four contemporaneous, or nearly contemporaneous, series of events'.[46]

Geological investigation thus resolved the earth's crust into two spatial series, one vertical and the other horizontal, that could be turned into a sort of chronological table. Each horizontal line of the table corresponded to the layers of minerals and fossils of the same epoch; Lyell distinguished four layers which he called Eocene, Miocene, old and new Pliocene. Vertically, four main classes of rocks 'form four parallel, or nearly parallel columns'.[47] Each body contained in the earth's crust can therefore be defined by two coordinates: one describing its epoch and its contemporaries, the other specifying the nature and even, in certain cases, the method of formation. The present corresponds to the bottom line of the table: it is the layer on the surface of the globe. At the other end of the table, in the part corresponding to the remotest times – that is, in the depths of the earth – there are no traces to be found to mark the beginning of the series, no signs to link the history of former ages with the presence of organized beings. Whether or not the remains found belong to what once lived, they always describe the same events of the same history. The living world is but one aspect of the earth and its past.

Evolution

In the history of nature, while fossils tell the age of the rocks, the layers of minerals, in return, describe the distribution in space and time of species that lived in the past. According to Lyell, this is how the geologist learns

what terrestrial, freshwater, and marine species co-existed at particular eras of the past; and having thus identified strata formed in seas with others which originated contemporaneously in inland lakes, we are then enabled to advance a step farther, and show that certain quadrupeds or aquatic plants, found in lacustrine formations, inhabited the globe at the same period when certain fish, reptiles and zoophytes lived in the Ocean.[48]

In view of the series of upheavals which occurred on the earth's surface, the same species have not been able to multiply and continue their existence throughout the entire history of the globe. On the horizontal plane of stratification of the earth's crust, the same types of organic debris are not spread over wide areas, but restricted to certain particular zones. Examining the distribution of organisms on the globe today shows that the habitable places on earth and in the sea are divided into a large number of different areas or 'provinces', each inhabited by a particular ensemble of animals and plants. In the same way, geology distributes living forms of ancient times over numerous provinces, each populated by a particular collection of animals and plants, provinces today buried in different zones at different levels. The succession of living beings on earth therefore appears to have developed, not by means of a single chain of transformations, but 'by the introduction into the earth from time to time of new plants and animals'.[49] To be able to grow and multiply during a certain period of time, such new organisms had to fit in with the conditions of life prevailing in that place and at that time.

Descending into the depths of the earth's crust is equivalent to going further back in time. According to Lyell's image, geological archives describe history in the globe, written in a changing dialect and incompletely preserved; only the last volume of history has been found, and its pages are in such poor condition that it is possible to decipher only fragments of each chapter, a few pages here and there. The vocabulary of this language, gradually changing from chapter to chapter, can in a way represent forms that once lived, then were buried in the consecutive formations of the earth and appear to have been flung into it. By comparing the last chapters, by attempting to read, from the order in which strata are superposed, the relative age of fossils they contain, new relationships between vanished families or species and those living today become apparent. According to Humboldt, 'All observations agree on the fact that the lower or more ancient the fossilized forms and flora, the greater difference they present with present-day animal or plant forms.'[50] In the world of living beings as in that of rocks, the present thus corresponds to the outermost layer of the earth's crust. Furthermore, numerical relationships can be established between what is now living and what once lived. A large number of present species is related to a small number of vanished species, as if the family ties between the past and the present could be represented by a cone with its apex driven far into the earth's crust. As if, starting with a single plan of organization and a single type, living bodies tended to diverge in the course of time.

The same sort of relationship is also revealed by the comparison of similar species, or even varieties of the same species living in different geological areas. For if the living world was formed by causes still at work today, it must be possible to see them in action under certain conditions. The investigations that led directly to the formulation of a theory of evolution had two facets: an inquiry into the geographical distribution of species; and a comprehensive analysis of the factors that intervene in the formation of species. With Darwin and Wallace, a new kind of naturalist appeared. No longer working in museums and zoological gardens, they were travellers, like the geologists, who examined their material on the spot. They went from island to island, from continent to continent, in order to study living organisms in their natural settings, and to compare their structures, their habitats and their behaviour. They accumulated observations, comparisons, measurements. They did not hesitate to experiment in the field, to submerge snails in the sea for fifteen days in order to ascertain their capacity for survival, for instance, and thus to determine the possibility of their transport from one land to another. The enormous mass of material they collected made it possible to analyse the relationships between living beings, their geographically determined variations, and their tendency to spread or to disappear. A whole network of interactions gradually emerged from the differences imposed on species by geographical conditions, by isolation and by the possibility of transfer from one place to another by sea or by air. So on oceanic islands no amphibians or mammals can be found, species occur there which exist nowhere else on earth, and a remarkable affinity is found between the species on such islands and on the nearest continent, although these species are not identical. In the Galapagos Islands, for instance, Darwin wrote, almost every product of the land and water bears the unmistakable stamp of the American continent. There are twenty-six land birds, and twenty-five of those are ranked . . . as distinct species, supposed to have been created here; yet the close affinity of most of these birds to American species in every character, in their habits, gestures and tones of voice, was manifest . . . it is obvious that the Galapagos Islands would be likely to receive colonists, whether by occasional means of transport or by formerly continuous land, from America.[51]

Each island has its own birds, as it were. But all these birds have a family likeness with each other and with those of the American continent. It is as if the differences stood out on a background of resemblances, as if these various species of birds were all derived from a common ancestor, and their individual characteristics were only the result of their isolation in their geographical territories.

Geographical inquiry thus reached the same conclusion as investigation of 'palaeontological archives': in the course of time, a small number of similar organisms produces a large number of different descendants. Now, the more these descendants diverge from the original type, the more they tend to live apart and reproduce among themselves, and consequently the more these differences are likely to be perpetuated. According to Darwin, varieties that have become very distinct from one another are eventually promoted 'to the rank of species'. In the end, two variables regulate the appearance of new species: the size of the population and the frequency with which differences arise between its members. First, the size of the population, because the largest groups multiply more readily: each major group tends to go on increasing, thus producing still more distinct characteristics. Secondly, differences between individuals, because the more the members of a group are diversified in shape, properties and habits, the more they are qualified to occupy varied habitats and adapt to them. The most numerous groups have the greatest possibility of producing varieties. The most distinct varieties have the greatest chance of finding new habitats, of settling in them and, by numerical increase, of supplanting the less distinct varieties. New varieties in a species enable it to exploit more effectively the resources of its environment, owing to a kind of division of labour in the area it occupies. In Darwin's words:

This tendency in the large groups to go on increasing in size and diverging in character, together with the almost inevitable contingency of much extinction, explains the arrangement of all the forms of life, in groups subordinate to groups, all within a few great classes, which we now see everywhere around us, and which has prevailed throughout all time.[52]

All this is expressed by the terms 'divergence', 'diversification' and 'dispersion'. The succession of living forms throughout time can no longer be represented by a table with a single column, or even with several parallel columns corresponding to independent series; the only figure capable of describing the diversification of a group is the genealogical tree.

As buds give rise by growth to fresh buds, and these, if vigorous, branch out and overtop on all sides many a feebler branch, so by generation I believe it has been with the great Tree of Life, which fills with its dead and broken branches the crust of the earth, and covers the surface with its ever branching and beautiful ramifications.[53]

The divergence of characteristics, together with the preservation of common features through heredity, explains the affinities linking all members of the same family or even of a higher group. Within one family derived from a common ancestor but divided into distinct groups, certain characters are transmitted which become progressively modified. All the species of this family are thus linked together by 'indirect lines of affinity of various lengths going back into the past through a large number of predecessors'. Proceeding in this way step by step, the whole living world can be arranged along the branches of the same genealogical tree. All beings, whether they lived long ago or are living today, can be derived from a very small number of ancestors or even from a single ancestor. The origin of beings, the 'dawn of life', fades into the depth of time. It can be guessed at, like the root of a tree, the point of the cone buried right down in the crust of the earth. It is reduced to a few organized beings which 'had only the most elementary shape' and in which all difference and peculiarity was erased.

The clear-cut divisions observed in the living world are due to the extinction of intermediate forms. 'Extinction has only separated groups: it has by no means made them', said Darwin.[54] Birds seem to be profoundly separated from other vertebrates, because a large number of forms connecting their ancestors to those of other forms have vanished. On the contrary, far fewer of the forms which linked fish to amphibians have disappeared: the division is, therefore, less abrupt. In the long run, only a small number of the oldest species was able to produce descendants. Since all those issued from the same initial species are considered to belong to the same class, the divisions of the major animal and vegetable kingdoms each contain a small number of classes. The genealogical tree weaves a network of kinship between all beings. Between two organisms, the connection is measured by the degree of cousinhood, as it were: 'All true classification is genealogical,' said Darwin; 'that community of descent is the hidden bond which naturalists have been unconsciously seeking, and not some unknown plan of creation, or the enunciation of general propositions, and the mere putting together and separating objects more or less alike.'[55]

The second part of the theory of evolution concerns the mechanisms operating in the variation of living beings, in the progression of their organization and adaptation. It is based on three principles. First, and in accordance with what Lyell had stressed, the causes governing the evolution of living beings in the past are indistinguishable from those at work today. There is no need to invoke exceptional phenomena for bringing the various living forms into being; no need to envisage upheavals that resulted in the simultaneous replacement of *all* beings. In Darwin's words, there is 'a striking parallelism in the laws of life throughout time and space: the laws governing the succession of forms in past times being nearly the same with those governing at the present time the differences in different areas'.[56] All forms living at present are the descendants of those who lived in former times. There is no doubt that the customary succession of generations was never interrupted and, therefore, that no universal cataclysm ever disrupted the entire world. All the changes occurred gradually without jolts. No new inventions appeared, only varieties differentiated by divergence and separation. The transformation of one species into another represents only the sum of small changes in a series of succeeding generations in the course of adaptation. Evolution progresses 'by little steps' and 'can never make a sudden jump'. The continuity of living beings was replaced by that of the slow, tenacious and irresistible growth of the genealogical tree.

As a result, the idea of necessity in the living world, of a harmony imposing a system of relationships between beings was definitively rejected. Palaeontological documents, geographical distribution of species, the development of embryos, the phenomena of divergent characters coming from a common ancestor, the increase of certain groups and the disappearance of others, all demonstrate the contingency of living beings and their formation. No preconceived plan, whether carried out at one stroke by a creation or progressively executed by successive transformations, can account either for the forms that have inhabited, and still inhabit the earth today, or for their distribution. According to Darwin, 'There is no sudden appearance of new organs which seem to have been specially created for some purpose.' The appearance of a new form is not inevitable. It is the result of numerous forces which combined at a certain epoch in a certain place. Had conditions been different, the living world would be different today or there might even be no living world at all. By driving out the devil of necessity, the theory of evolution freed the living world from all transcendency, from any factor whose cause lay beyond knowledge. Nothing remained which, by its essence, blocked investigation and experiment.

Finally, the most radical transformation of the biological attitude wrought by Darwin was to focus attention, not on individual organisms, but on large populations. Until then, variations that could occur in a particular organism provided the yardstick for envisaging the types of transformations to which it might eventually be susceptible. With Darwin, the mishaps and misadventures that might happen to this or that individual lose all interest. Obviously, there is no chance of ever finding the trace of every animal that has once lived on earth. But even if the individual destiny of every past organism could be reconstructed, it would still be impossible to derive therefrom the laws of variation and evolution. The object of transformation is not one organism, but the ensemble of similar organisms who live in the course of time. Throughout his writings, Darwin continually stressed the abundant production of organized beings, the extent of destruction, the ineffectiveness of the mechanisms governing fertilization and reproduction. For among the millions of germ cells produced, it is the exceptional one that manages to play its role. Everywhere emphasis is put on this prodigious wastefulness of nature, through which the most infrequent events end up by having the most important consequences. Before Boltzmann and Gibbs, Darwin had already adopted, in a biological context, the outlook which statistical mechanics was to impose, in the second half of the nineteenth century, on the study of the behaviour of molecules. The whole theory of evolution is based on the laws of large numbers. Not that Darwin turned to complex mathematical methods for investigating population variations: he relied on intuition and common sense. In envisaging transformations, he considered only the fluctuations which always occur in large populations, what statisticians call 'tails of distribution curves'. His approach was already that of statistical analysis: the small advantage that a slight increase in the chances of survival and reproduction confer on a few individuals is transformed into a rigid mechanism with inevitable consequences. Necessity has not completely disappeared from the living world. It has merely changed its nature.

The main impetus for transforming living forms must, therefore, lie in the very process that gives rise to large populations – that is to say, the power of multiplication peculiar to living beings. There is no longer a simple, unique and continuous impulse to produce new forms in the course of time. The emergence of organisms represents the consequence of a long struggle between opposing actions, the resultant of contending forces, the outcome of a conflict between the organism and its environment. But in this respect, the organism always has the first word. Environment only replies. For Darwin, it is 'unreasonable' to believe that the feet of geese have become webbed simply by striking the water; 'absurd' to think that reptiles have lost their feet merely by trying to crawl better. The capacity of living beings to modify their forms, properties and habits is inherent in life itself. It is one of the qualities by which living beings differ from inanimate things. It is indissolubly linked to the most characteristic ability of organisms: that of reproduction. It is simply the expression of the old adage 'be fruitful and multiply'. By its very nature, each organism or each pair of organisms is endowed with the power of producing offspring in ever increasing numbers from generation to generation. If only one species existed on earth, with nothing to limit its expansion and nothing to destroy it, it would multiply indefinitely in geometrical progression. 'Even the least prolific of animals would increase rapidly if unchecked,' wrote Wallace.[57] Linnaeus had already calculated that if an annual plant and each of its descendants produced only two seeds a year – and there is no plant so unproductive as this – then in twenty years there would be more than a million plants. Darwin made the same calculation for the elephant, reckoned to be the slowest breeder of all known animals: assuming that it procreates only when thirty years old, that it lives a hundred years and that it brings forth six offspring in this interval, then the descendants of a single pair of elephants would total about nineteen millions in 750 years. As for man, who doubles in number in twenty-five years, after a thousand years at this rate, there simply would not be enough room for all the descendants of a single couple to stand side by side on the surface of the earth. But no single species exists alone on earth. What exists at any time are populations of different organisms, in competition with one another for territory, food and light; in short, for their very existence. The effect of environment, then, is limited to favouring the multiplication of some species at the expense of others. Some are doomed to die out, others to expand. Now, if all species are able to multiply, if all can produce more organisms at an ever increasing rate, generally these descendants are almost, but not quite, identical to the initial type. As soon as they appear, varieties take part in the competition. They win or lose, depending on whether or not the difference between them and their ancestor favours multiplication. Hence the gradual replacement of certain species by others better adapted to reproduce under certain conditions. In the end, the only force peculiar to the evolution of the living world is the power of multiplication peculiar to living beings.

How wide a gap separates this theory from all that preceded it! The only exception was Malthus, to whom both Darwin and Wallace attributed the idea of population equilibria and the notion of conflict between two opposing forces – one inherent in the organisms themselves and the other external to them – as a regulating factor. To build a theory of evolution, biology thus borrowed a model from sociology. But this model was itself based on the stability that Buffon had already observed in the number of individuals in one species. 'The cause to which I allude is the constant tendency in all animated life to increase beyond the nourishment prepared for it,' said Malthus . . . 'Through the animal and vegetable kingdoms Nature has scattered the seeds of life abroad with the most profuse and liberal hand; but has been comparatively sparing in the room and the nourishment necessary to rear them.'[58] For Malthus, the development of human populations was subject to the action of two factors working in opposite ways: on the one hand, 'multiplication by geometrical progression'; on the other, external obstacles such as destruction, war, scourges, famine; in short all the restrictions necessarily brought to bear on expansion, particularly the fact that the 'means of existence' themselves cannot increase at the same rate, but at best by arithmetical progression. Hence a conflict that, in human societies, produces a 'struggle for existence' in which Malthus saw both the cause and the consequences of the social changes that arose with the advent of the industrial age. When the theory of evolution adopted the idea of a struggle for existence, however, it was in a slightly different sense. By a curious misconception, biological evolution has often been cited as a prime example of vital competition, the victory of the strong over the weak, of masters over slaves, in order to find a natural basis for social or racial inequality and to justify their worst excesses. In fact, Wallace, and even more Darwin, mainly borrowed from the doctrine of Malthus the idea of an interaction between the power of reproduction and the outside forces limiting that power.

Like Maupertuis before him, Darwin strove to find in the process

of artificial selection carried out on domestic animals and cultivated plants a model that could apply to nature. Now, the production of new species by breeders and horticulturists results from an
interaction of two kinds of events: on the one hand, living beings
tend to reproduce and diversify naturally; on the other, breeders
endeavour to develop, in their herds or crops, those varieties of particular interest to them. It is not man who acts directly on variability.
Modifications appear only 'occasionally' without our knowing how
to produce them. They arise spontaneously, as it were, unconnected
with the slightest need or the least requirement of the organism. Thus,
according to Darwin, 'The chance of their appearance will be much
increased by a large number of individuals being kept; and hence
this comes to be of the highest importance to success.'[59] But although
it is nature and not the breeder who produces variability, in return
the breeder can *choose* from the variations provided by nature and
adapt animals to his own design. Once a variation has appeared in
an individual, it is maintained by heredity in its offspring. Mere
individual differences are therefore enough to allow changes to
accumulate in the desired direction. Man can make a selection, either
methodically or even unconsciously, by keeping the most useful or
most pleasing individuals at each generation. By placing organisms
under the desired conditions, by regularly sorting out individual
differences – sometimes so slight as to escape an untrained eye – and
by allowing only these organisms to reproduce, any characteristic
whatsoever of an animal or a plant can be modified.

For Darwin, this model of artificial selection can be applied in the
form of 'natural selection' to the phenomena of variation and
evolution in nature.

This preservation of favourable variations and the rejection of injurious variations, I call Natural Selection, or Survival of the Fittest.
Variations neither useful nor injurious would not be affected by natural
selection and would be left a fluctuating element.[60]

Variability occurs spontaneously in the course of generations, in
nature as in stock-breeding. In both cases, the size of the populations
plays a major role in providing the opportunities for variations to

appear. This is where the time factor comes in. In itself, duration has no influence on natural selection.

Lapse of time is only so far important, and its importance in this respect is great, that it gives a better chance of beneficial variations arising, being selected, increased, and fixed, in relation to the slowly changing organic and inorganic conditions of life.[61]

Never before had time been measured by generations. But, in fact, the generation represents the most suitable unit for estimating the time required for new forms to appear during geological periods. In the genealogical tree Darwin drew to represent the living world, a new branch appears when the extent of accumulated variation is sufficient to characterize a well defined variety, such as one recorded in a book of systematic zoology. The intervals between two branches 'may represent each a thousand generations or ten thousand'.[62]

The processes of natural selection resemble those of artificial selection, but their effects are much more widespread. Whereas the breeder can choose only visible characteristics to suit his taste or needs, nature can also act on internal organs, on the very constitution of living beings and on the whole organization.

It may be said that natural selection is daily and hourly scrutinising, throughout the world, the slightest variations; rejecting those that are bad, preserving and adding up all that are good; silently and invisibly working, whenever and wherever opportunity offers, at the improvement of each organic being in relation to its organic and inorganic conditions of life.[63]

The struggle for survival is above all a struggle for reproduction. Individuals are constantly and automatically tested for their ability to multiply under certain conditions of existence and produce descendants that can live in certain territories. Selection operates even in sexual phenomena: the struggle between males for the possession of females results in the strongest and most wily having the most descendants.

When the males and females of any animal have the same general habits of life, but differ in structure, colour and ornament, such differences have been mainly caused by sexual selection: that is, individual males have had, in successive generations, some slight advantage over other males, in their weapons, means of defence, or charms; and have transmitted these advantages to their male offspring.[64]

The most trifling advantage an organism may have over its rivals of the same species tips the scales in its favour. The slightest change in efficiency of reproduction is enough to change the population balance. Some modifications enable a few individuals to multiply rather better and faster than others. These few tend to increase, and the others to disappear. Since most variations are transmitted by heredity, beneficial changes accumulate naturally from generation to generation, while the others are eliminated. According to Wallace,

All variations from the typical form have some definite effect, however slight, on the habits or capacities of the individuals . . . If any species should produce a variety having slightly increased powers of preserving existence, that variety must inevitably in time acquire a superiority in numbers.[65]

Today as yesterday, evolution is at work to maintain, correct and improve the adaptation of animals and plants to their environment. Natural selection operates by means of differential reproduction.

*

Until the middle of the nineteenth century, the living world was considered as a system controlled from the outside. Whether they had been fixed since creation or whether they had progressed through successive events, organized beings were always arranged in a continuous series of forms. Apparent breaks in the hierarchy were due to omission, ignorance or inadequate listing. The existing structure of the living world, then, expressed a transcendental necessity. That living beings could be any different from what they are, that other forms might inhabit the earth, this was just inconceivable. The theory of evolution swept away the idea of a preconceived harmony that imposed a system of relationships on organized beings. The necessity for the living world to conform to its present pattern was replaced by the contingency that already governed the sky and inanimate things. Not only might the living world have been totally different; it might equally well never have existed at all. Organisms became components of a vast system of higher order embracing the earth and everything on it. The form, properties and characters of living beings, therefore, were subject to regulation from within the system – that is, to the interactions coordinating the activity of its components.

This revolution in outlook was not merely a continuation of the transformist way of thought that originated with Buffon and Lamarck. It reflected a change in the very way of looking at objects; it was the result of a radically new attitude that appeared in the middle of the nineteenth century. The evidence that such a transformation was not a mere accident comes from its independent and more or less simultaneous occurrence in very different fields: in the study of matter, with Boltzmann and Gibbs, as well as in the study of living beings with Darwin, Wallace and a little later, Mendel. There are, indeed, two ways of looking at a collection of objects of the same kind, such as molecules of a gas or organisms of the same species. On the one hand, they can be considered as a group of identical bodies, all members of the group being true copies of the same pattern. In the living world, the forms are thus classified according to the structure perpetuated through successive generations, i.e. the permanence of the type. It is not the objects themselves that have to be known, but the type they represent. Only the type has any reality. Objects merely reflect the type. It does not matter if the copies are sometimes different from the pattern. Deviations from the type are negligible quantities or insignificant defects. On the other hand, the same collection of objects can be seen as a group of individuals which are never identical. Each member of the group is unique. There is no longer a pattern to which all individuals conform, but a composite picture, which merely summarizes the average of each individual's properties. What has to be known, then, is the population and its distribution as a whole. The average type is just an abstraction. Only individuals, with their particularities, differences and variations, have reality.

These two ways of looking at nature and its objects thus clash in every respect. It was the passage from the first to the second that marked the beginning of modern scientific thought. In the study of the inanimate world, the new attitude found expression in the appearance of statistical mechanics. In the living world, it was a necessary pre-condition for a theory of evolution. Variation was no longer a matter of individuals, but of populations. Even though Darwin did not use statistical analysis, he had a statistical conception of populations. First, because variations only express the fluctuations of distributions inherent in every system; secondly, because selection acts only by slowly altering population equilibria through the random interaction of organisms and their environment. This new outlook eliminated the main difficulty hitherto encountered in explaining how transformations occurred in living beings. There was no need to refer to some complicated mechanism, to some design by nature or to some environmental influence to explain changes in form. No individual of any species was any longer identical to its fellows. In each generation, each character went through the complete series of slight deviations from a mean. Only later did adaptation come into play by virtue of the fact that any organism appearing on earth is immediately put to the test of life and reproduction.

The theory of evolution is therefore characterized by a way of considering the emergence of living beings and their aptitude to live or adapt themselves to their environment. For Lamarck, when a new organism was formed, its place was *already* marked out in the ascending chain of beings. It had *in advance* to represent an improvement, a progression on what had previously existed. Trend, if not intention, came before realization. With Darwin, this order was reversed: the formation of an organism precedes its adaptation. Nature only favours what already exists. Production comes before any value judgement on what has been produced. Any change whatsoever can result from reproduction. Any variation may appear, whether an improvement or a regression in comparison with what already existed. There is no manichaeism in the way nature produces novelties, nothing of progress and regression, of good and evil, of better and worse. Variation occurs at random, that is, without any relation between cause and result. Only after a new being has emerged is it confronted by the conditions of existence. Only once they exist are candidates put to the test of reproduction.

In the eighteenth century, Buffon, Maupertuis and Diderot had already envisaged the possibility of organisms being sorted out after they had been formed. Monsters could be produced, which unable to live, had to disappear. Nature was screening, in a way, among already formed organisms, keeping only those equipped with the means of living and rejecting what was incomplete, unable to feed for lack of mouth or to reproduce for lack of genital organs. This type of monster, however, could only be imagined in terms of the interplay of organic molecules. All combinations possible *a priori* would materialize, but only some could be viable. In the middle of the nineteenth century, the situation was completely different. The equilibrium of the living world was established through a kind of dialectic of permanence and variation, of identity and difference. Darwin's application of the term 'selection' to evolution, by analogy with breeding, has sometimes been criticized, since the idea of choice is associated with the intention that leads the breeder to sort from his stock the forms apparently best suited for his purpose. There is also a sorting operation in nature; but it is automatic. Everything that can interfere with reproduction in some way modifies it. Among the candidates for reproduction, no intention directs the choice of the elect, which is made *a posteriori* on the testing bench, and results only from the qualities and performances of the individuals. Adaptation becomes the outcome of a subtle interplay between organisms and their environment. For although the power to reproduce is inherent in the organism, its realization closely depends on all the variables of environment. An organism 'chooses' its environment as much as the environment chooses the organism. In the process of adaptation, reproduction operates only as an amplifier. It merely accentuates deviations that occur spontaneously. By always pulling in one direction, reproduction can in the end make populations drift along well defined paths. Occurring blindly, variations take the direction imposed by the inexorable sorting process of natural selection. For Darwin, a living being becomes, from the time of its birth, part of the immense organized system formed by the earth and everything on it. Natural selection represents a regulatory factor that maintains the harmony of the system. Today we consider that a system of this kind can survive only if the 'feed-back' loops automatically adjust its functioning. Evolution thus becomes the result of feed-back from environment to reproduction.

Natural selection acts slowly and in stages. Evolutionary time is irreversible, unlike the time that was still operating in physics. In Newtonian mechanics, in fact, there was no preferential direction in the interaction of two bodies. It was statistical thermodynamics in the second part of the century that introduced the concept of time's arrow into physics, by making populations of molecules evolve from the least probable towards the most probable state – from order to disorder. With the theory of evolution, however, time in the living world was already unidirectional: once living beings are driven along a certain path by variation and selection, they cannot turn back. Natural selection compels them either to continue differentiating in the way indicated or to disappear. According to Darwin, under the conditions imposed by life on earth, the result of natural selection 'is that each creature tends to become more and more improved in relation to its conditions. The improvement inevitably leads to the gradual advancement of the organization of the greatest number of living beings throughout the world.'[66] At each stage in the series, there is only a very slight probability of returning to an earlier state. The chance of ever retracing the complete path is therefore nil. On this principle can be based a general theory of the evolution of organized systems, whether living or not. Increasing complexity in the course of time and the irreversibility of the sequences of transformations become properties inherent in such systems. What is called 'progress' or 'adaptation' is only the necessary result of the inevitable interplay between the system and its surroundings.

In the middle of the nineteenth century, ideas about heredity were still very vague. But in the theory of evolution, what was selected acquired the permanence of heredity. Although not formulated by Darwin, this proposition is implicit in his work. Reproduction, therefore, shapes both identical and different forms. Its regularity moulds a child in the image of its parents. Its fluctuations create novelties. Living beings are born, with or without modifications.

Then they are judged – judged by the country they live in and the organisms around them; by those they hunt and those that hunt them; by their own sex and by the opposite sex. The verdict is final and without appeal; it is measured by the number of descendants. It is easy to understand why reproduction was thenceforth given cardinal importance. It became the main factor operating in the living world, the source of both stability and variation, the process by which structures, qualities and attributes of living beings were maintained and diversified. Reproduction was the meeting point of determinism, responsible for the formation of like forms, and contingency, responsible for the appearance of new forms. For, with the theory of evolution, necessity changed both its nature and its target in the living world. When applied to the behaviour of enormous populations, it became what statistical analysis was to consider as the expression of laws governing large numbers. But it no longer reflected an incomprehensible force that imposed the present pattern of the living world. As long as living forms were all linked by an *a priori* system of necessary relationships, a whole sector of the living world was excluded by essence from investigation and experimentation. Faced with this inaccessible realm, the biologist was somewhat like a child with his nose pressed against the window of the cake-shop, gazing at forbidden delights. But once necessity was confined to the mere effects of a selection resulting from the obligation to live under certain conditions, in certain areas and among certain beings, the inaccessible vanished. If no intention was responsible for the appearance of new forms, then their success or failure in the 'struggle for existence' depended only on physical factors, that is to say, variable parameters. No sphere of biology remained inaccessible. Even reproduction could become the object of investigation.

4

The Gene

The middle of the nineteenth century marked a turning-point in biology. In less than twenty years, there appeared the cell theory in its final form, the theory of evolution, the chemical analysis of major physiological functions, the study of heredity, the study of fermentation, and the first total synthesis of organic compounds. Through the work of Virchow, Darwin, Claude Bernard, Mendel, Pasteur and Berthelot were defined the concepts, methods and objects of analysis that form the basis of modern biology, for the approach then adopted has scarcely changed during the twentieth century. Hitherto confined to observation alone, biology grew into an experimental science. During the first half of the nineteenth century, organization had formed the basic characteristic of all living beings. It represented the second-order structure that controlled everything perceptible in the organism. Placed in the very core of each living being, it served as a fulcrum, the master plan to which all observation and comparison of the visible structure and properties of living beings were referred. In the second half of the nineteenth century, in contrast, organization no longer remained the starting point for understanding living beings; it became what had to be known. It was no longer enough to state that organization underlay all characteristics of an organism. It was also necessary to discover at every level what it was based on, how it became established, what laws controlled its formation and functioning. This change in the approach to organization revealed a whole set of new possibilities for investigation. Thereafter biologists no longer examined life as a hidden, irreducible and inaccessible force emerging from the depths of time. They looked at those factors into which life had been decomposed: its history and origin; causality, chance and function. To the organism as a whole, new objects for investigation were added: cells, reactions and particles.

Biology then divided into two branches, each with its own techniques and material. On the one hand, some biologists continued to study the organism as a whole, either as a unit or as part of a population or a species. This form of biology, which had no contact with the other natural sciences, operated with the concepts of natural history. The habits, development, evolution and relations of animals with other species could thus be described without any reference to physics or chemistry. On the other hand, some biologists attempted to reduce the organism to its constituent parts. Physiology required it; the time was ripe. All nature had become history; but a history in which beings were the extensions of things, in which man fell into line with animals. The introduction of contingency in the living world by Darwin and Wallace corresponded in biology to the 'everything is permissible' of Ivan Karamazov. No sphere in the study of living organisms remained inaccessible; no area was in principle beyond the reach of knowledge; divine law no longer set limits to experimentation. In a universe where creation had been replaced by contingency, the ambition of biology became unlimited. If the living world was progressing at random, if it was shorn of finality, then it rested with man to master nature. It rested with him to establish the order and unity that he had previously sought in the essence of life; and all the more as dialectics and positivism were trying to re-build the bridge between the organic and the inorganic that had been destroyed at the end of the eighteenth century. Between inert matter and living beings, the difference was not of nature but of complexity. The cell was to the molecule what the molecule was to the atom; a higher level of integration. To study this kind of biology, it was no longer sufficient merely to observe living beings. It was necessary to analyse their chemical reactions, to study their cells and to reveal new phenomena. If the organism still had to be considered as a whole, it was because the regulation of reactions, the coordination of cells and the integration of phenomena made a synthesis possible.

At the end of the nineteenth and the beginning of the twentieth century, a whole series of new objects became accessible to study. Around each of these, a particular field of biology developed. This science gradually became divided into compartments; the word 'biology' came to mean a whole range of different sciences, widely separated not only in aims and techniques, but even in materials and vocabulary. Two of the branches that came into being at the beginning of the twentieth century completely refashioned the existing picture of organisms, of their function and evolution: biochemistry and genetics, each embodying one of the tendencies of biology. Working with extracts, biochemistry studies the components of living beings and the reactions that take place in them; it refers the structure and properties of living beings to the network of chemical reactions and to the behaviour of a few species of molecules. Genetics, in contrast, examines populations of organisms to investigate their heredity; it refers both the production of identical forms and the difference of novelties to the properties of a new structure hidden in the nucleus of the cell. While itself obeying strict laws, this third-order structure governs the characteristics and activities of the organism at all levels. It controls the development of the embryo. It determines the organization, form and essential characteristics of the adult. It maintains species in the succession of generations and produces new species. This structure houses the 'memory' of heredity.

Experimentation

Until the middle of the nineteenth century, biology remained largely an observational science. Properties and structures were defined in terms of the whole organism; living bodies were compared with each other to determine the analogies and differences. For Darwin as for Cuvier, it was nature that carried out experiments for the naturalist. When anatomists wanted to determine the internal structure of organs, they opened cadavers. When histologists tried to resolve animals and plants into their elementary components, they examined tissues under the microscope. When embryologists studied the development of the egg, they watched cells dividing, layers forming and organs developing. Only physiologists sometimes attempted deliberately to modify the conditions of life and to observe the results. Even they, however, did not work with isolated organs or tissues but with the whole organism. Since the time of Lavoisier, the necessity of a close association between chemistry and physiology had been obvious. These two branches of science, however, could not agree either on methods or on materials. The necessity for postulating a vital force to justify the molecular properties of organisms raised, between the chemistry of life and that of the laboratory, an insuperable barrier.

After the middle of the nineteenth century, it was no longer enough simply to refer the structure of organs to their functions in order to determine the relationships between them. The actual operation of living bodies and of their constituent parts had to be investigated. That was where physiology came to the forefront, but its nature changed. In Cuvier's day, physiology represented above all a system of reference for anatomy; it helped to establish the analogies on which the comparison between living beings and their organization was based. To Claude Bernard, physiology appeared in a completely different light. The functioning of an organ could no longer be interpreted in terms of structure and texture. It had to be analysed, to be resolved into different parameters and even, as far as possible, to be measured. It was anatomy that became the auxiliary of physiology; no longer a physiology of observation, involving what Claude Bernard called 'passive experimentation' – in which the biologist simply notes the variations occurring spontaneously in a system – but an 'active' science, in which the experimenter intervenes directly, chooses an organ, isolates it, makes it function, modifies the conditions and analyses variables. Biology had to change its place of work. Previously, it had been conducted in nature: when the naturalist was not in the field observing living beings in their own setting, he was working in a natural history museum, a zoo or a botanical garden. Thenceforth, biology was carried out in the laboratory.

There were at least two reasons for the attempt to investigate the function of the living organism not as a whole, as had been done in the past, but in separate parts. Firstly, in the middle of the century the necessity of invoking a vital force became less stringent. Since the work of Bichat, each living being was considered the field of battle between the forces of life and those of death, between production, which was under the influence of some factor peculiar to the living, and destruction, which resulted from physical and chemical activities. With the development of thermodynamics and with the total synthesis of organic compounds, the barrier erected between the chemistry of life and that of inanimate matter collapsed. Secondly, the cell theory presented living beings, not as indivisible entities, but as associations of components. However complex an organism appeared, it was never more than the sum of its elemental units. 'In the last analysis,' said Claude Bernard, 'it is an edifice of anatomical elements, each with its own existence, evolution, beginning and end; total life is only the sum of these individual lives, associated and harmonized.'[1] Physiology had to be studied, therefore, by resolving the complexity in a completely Cartesian manner. Whenever possible the individual components should be studied, and not the organism as a whole. In this respect, physiology had to adopt the approach of other experimental sciences. For Claude Bernard,

Just as physics and chemistry discover the mineral components of compound bodies by experimental investigation, so to comprehend the phenomena of life, that are so complex, it is necessary to go deep into the organism and to analyse the organs and tissues in order to reach the organic components.[2]

When an animal breathes, it is the red blood corpuscles and the cells of the lung that are working; when it moves, it is the fibres of the muscles and nerves; when it secretes, the cells of the glands. Organs and systems exist, not for themselves, but for the cells that build up the structures and perform the functions. Their role is to provide in quality and quantity the conditions required for cellular life. Vessels, nerves and all the different organs are arranged so as to create a suitable environment around each cell and to provide it with the appropriate materials, food, water, air and heat. In an organism, therefore,

The cell component is autonomous since it has in itself and as the consequences of its protoplasmic nature the conditions essential for its life, neither borrowing nor appropriating from its neighbours or from the whole organism; on the other hand, it is linked to the whole organism by its function or the product of that function.[3]

To describe the living organism, Claude Bernard used the metaphor of societies or factories in which, by division of labour, all elements work in the common interest. The organs 'are in the living body what factories or industrial establishments are in an advanced society, that provide the members of the society with the means of clothing, heating, feeding and lighting themselves'. Studying physiology means investigating such a system.

The very complexity of living bodies raises, however, two kinds of difficulties. First, in attempting to reach the most deeply hidden components of the organism there is a risk of seriously injuring it, disturbing or even preventing its functions. Experimentation in the organism, therefore, has to be introduced gradually, studying first the major functional systems, then organs, then tissues and only at the end the cells that contain the properties of life. The second difficulty is that, in living beings, the phenomena which take place in the various organs are not independent of each other. In plants or lower animals such as Hydra or Planaria, fragments cut out of the organism are able to survive and grow. In higher animals, however, it is the subordination of the parts to the whole that makes the organism a united system, an individual. Although each cell has the properties of life, although it leads a more or less autonomous existence, it nonetheless works for the community. The physiologist, therefore, must try to break down the organism and isolate its components by experimental investigation, but without thinking of them separately. The physiology of an organ can be interpreted only by reference to the whole organism. 'Determinism in the phenomena of life is not only very complex, but also harmoniously subordinate,' said Claude Bernard.[4] It is neither their nature nor a property peculiar to the living that makes biological phenomena more complicated than those of physics. They are more complicated because they can never be isolated. They are always the result of a series of events that remain indissolubly linked and generate one another. In physiology, the interaction of functions and their mutual dependence create complexity.

Far from attempting to exempt living beings from the laws controlling inanimate matter, the physiologist must, therefore, try to analyse by physical and chemical methods the phenomena that take place in the organism. Not that physics and chemistry can resolve all biological problems, but because they are simpler than physiology, and because the simple must always be used to clarify the complex. Biology ought to 'borrow the experimental method from physics and chemistry but retain its own special phenomena and laws', said Claude Bernard. Then physiology could be transformed into an active science. Until that time, interfering with an organism could only impair its nature and damage its functions. Thenceforth, it became possible to operate on a living body and to carry out experiments without necessarily destroying the very quality of life in the artificial conditions so established. It became possible to separate certain constituents of the body by mechanical methods, to study their functioning and, with certain precautions, to draw conclusions as to their natural role in the organism. It was important to untangle the network of operations that take place simultaneously in a living body, to experiment under conditions as well defined as possible and to isolate simple phenomena. According to Claude Bernard, the physiologist must be 'an inventor, a true foreman of creation'.[5]

A phenomenon can be born by chance observation or as the logical result of an hypothesis. Actually, there are two almost infallible recipes for manufacturing phenomena. The first is to reproduce experimentally what nature brings about by disease. In a sense, physiology and medicine represent two branches of the same science, not only because of the object they study, but also because of the similar methods they use. In the relationship between the normal and the pathological, each discipline guides the other. Medicine can no longer remain empirical, but has to be based on the results of physiological investigation. Conversely, knowledge of pathological conditions contributes to the knowledge of physiological conditions. Medicine opens up the path for physiology. It shows where to act. It indicates the effects to be achieved. A machine can be repaired only if its parts and uses are understood. In turn, deliberately damaging an organ enables its role to be defined. Pathology, therefore, suggests models to the physiologist, who tries to reproduce disease by provoking lesions in as limited an area as possible and to analyse the consequences. Deliberately and specifically damaging a component of the body by mechanical or chemical means, determining the effects of the lesion and defining the reactions of the other components is one of the most effective methods available to physiology. One can remove an organ, like the kidney; or destroy it *in situ*, like the pancreas by injection of paraffin; or injure it, like the gall bladder after ligature of the bile duct. By comparison with a healthy animal, the effects of the lesion can be deduced. One can even try to treat such injured animals by administering certain substances, or even undamaged extracts of the affected tissues. One can also try to compensate for certain disorders by making further lesions in other organs. This method of exploration by mechanical lesions opened up the way for analysis of many functions. For instance, it led to the distinction between 'excretion' by a tissue that produces nothing in itself but allows evacuation of substances formed within, and 'secretion' by a gland that attracts certain compounds, combines them and creates new substances. The secretions themselves could be classified in two categories: 'external' when the product is secreted outside the organism, and 'internal' when it flows into the organism to help digestion or some other function. There was a century of exploration ahead in this field. It was, in fact, the method later used to define certain metabolic processes in higher animals, to analyse digestion, discover the existence of hormones, determine the role of certain nerves and locate the functions of the brain.

Mechanical interventions can be replaced by chemical lesions. The effects of poisons simulate diseases. Moreover, they act electively on organs. After injection of a toxic product, lesions can almost always be seen in a particular tissue, wherever it occurs in the body. It is always a given histological component that is affected by a given poison. Carbon monoxide, for instance, attacks red blood corpuscles and prevents them from playing their part in respiration; certain metallic salts damage cells of the kidney, hindering the evacuation of the waste-matter from the blood via the urine; curare affects nerve cells, stopping the transmission of the nervous impulses and thus paralysing the animal. It is even possible to look for antidotes which specifically neutralize the action of a poison and allow the affected organ to function again. With the arsenal of poisons, physiology has an unequalled instrument at its disposal: simplicity of use, specificity in action, control of the effect by regulating the dose, sometimes even reversibility of the lesions. The use of toxic agents was, therefore, to become one of the methods most favoured by physiologists for more than a century. It still plays an important role in the investigation of functions and in the study of chemical reactions, whether in the organism, in the cell or in cell-free extracts.

The second recipe for producing phenomena in physiology is based on the equilibrium, ever unstable but ever-recurring, between the organism and its environment. This interaction is so close that the organism is, as it were, obliged to react to every change in environment. According to Claude Bernard, living and inanimate bodies are alike in this respect. There are always two things to be considered in every phenomenon: the object under observation and the external circumstances which affect this object and lead it to reveal its properties. Were the environmental factor eliminated, the phenomenon would disappear, as though the object had been removed. There is attraction only to the extent that the behaviour of two bodies is observed; just as there is electric current only through the relation between copper and zinc. Let one of the bodies or the copper be removed, and neither attraction nor electric current will be found: they become abstract ideas. The same is true of living beings. 'The vital phenomenon is therefore neither wholly in the organism nor in its surrounding,' said Claude Bernard; 'it is in a sense the effect produced by the contrast between the living organism and its environment.'[6] Eliminating or corrupting the environment amounts to removing or destroying the organism. In the last extreme, the organism can even be considered as a reagent of its environment, as Lavoisier had already done. The environment is no longer just the fluid, air or water in which the organism is immersed, however. With Auguste Comte, it also became heat, pressure, electricity, light, humidity, oxygen or carbon-dioxide content and the presence of beneficial or toxic chemical compounds – in short, everything that comes into contact with the envelope of a living organism and acts on it in some way. Since each of these factors can be modified, each becomes a parameter for experimentation. In the organism–environment system, two series of variables are coupled: some external, on which experimentation acts at will by physical and chemical means; and the others internal, expressed in the functions which are also to be measured by physical and chemical methods. Manipulating the former makes it possible to reach the latter. To produce a phenomenon, it is sufficient to place a living being, an organ, a piece of tissue or even an extract under environmental conditions as well defined as possible, and systematically to vary each parameter of the environment. It is no exaggeration to say that since then this procedure has constituted the principal activity of biological laboratories.

Until then, living beings had been classified according to their degree of complexity. A functional relationship was now added to the structural one. Correlation can be established in living beings between the degree of organization and the kind of interaction with the environment. On the one hand, there are the simple organisms reduced to a single anatomical component, such as infusoria, or even composed of several cells, such as the lower animals and plants. In these organisms, all the component parts come into direct contact with the surrounding medium – that is to say, the water or air around them. On the other hand, there are more complex organisms, in particular the higher animals, composed of a large number of cells. There, only the surface components are in direct contact with what Claude Bernard called the 'cosmic medium'. The parts hidden in the depth of the organism are immersed in an 'internal' or 'organic medium' which acts as an intermediary with the cosmic medium. In man, for instance, the essential components, those performing the most important functions, are not exposed to the variations and hazards of the external medium. They are in contact only with the blood and body-fluids that protect them from any sudden change. For the chief characteristic of the internal medium is its constancy. Produced by organs for organs, it acts as a shock-absorber, a buffer to protect the most valuable elements of the body from inopportune changes, so they can work under almost invariable conditions. According to Claude Bernard,

It is therefore quite true to say that the fowls of the air do not really live in the atmosphere, or a fish in water or an earth-worm in the soil. The atmosphere, water and earth are a second envelope around the substratum of life, that is already protected by blood circulating everywhere and forming a first protective fence around all living particles.[7]

Higher animals literally live within themselves.

The concept of an internal medium justifies in terms of function the way Cuvier had distributed the structures of an organism in space. The most valuable components are buried deep in the body, because they are thus better protected from misadventures in the surrounding medium. So they can work, sheltered from slight variations in temperature, humidity, pressure, etc. Complexity of organization means freedom of functioning. And independence from the environment becomes in the end a selective factor in evolution. It is thus possible to turn the classification of living beings upside down: they can be redistributed according to the nature of their 'medium', that is to say, their autonomy with regard to the outside world. There are three kinds of existence. In a first group, which comprises the lower beings, there is total dependence on external conditions. When conditions are right, life follows its course. If they become unfavourable, the organism dies or falls into a 'dormant state' of 'chemical indifference': all exchanges, all activities, all manifestations of life are suspended. In a second group, which includes lower animals and plants, the internal medium is less closely dependent on external conditions; so that fluctuations in environment still affect the life of the organism, decreasing or increasing its vitality but never suppressing it entirely. It is generally the temperature of the body that, under the influence of the external temperature, regulates the movements of this 'fluctuating life'. In the third and final group, that of higher animals, all activities are entirely independent of external conditions. Whatever the vicissitudes of their surroundings, organisms go on living in the same way, as if in a hothouse: it is the 'constant and free life, independent of variations of the cosmic medium'. The more complex the organism, the more independent it is.

The internal medium and its stability emphasized one of the fundamental properties of living beings: the regulation of functions. They provided a way of measuring the degree of integration of an organism by allowing experimentation on the coordination between organs and functions. This investigation led to representing the organism as a system in which all activities are adjusted right down to the last detail. Already in the eighteenth century, an interaction of certain functions had been considered. For Lavoisier, the animal machine was commanded by three principal 'regulators': respiration, digestion and transpiration. For Claude Bernard, it was all the activities of the body that were subject to regulatory mechanisms. 'All vital mechanisms, however varied, have but one objective, that of maintaining the unity of life functions in the internal medium.'[8] There were mechanisms of 'equilibration'; there were 'compensatory', 'insulating' and 'protective' mechanisms regulating temperature, concentration of water, oxygen content, food storage, the composition of the blood and external or internal secretions. The more complex the organization of an animal, the more precisely the regulatory systems were adjusted. Hence the advantage in physiology in using organisms in which these systems were most perfected, that is, the higher animals. For biology in the nineteenth century, the organization of living beings was based primarily on the integration of functions. According to Claude Bernard, life is possible only because 'there is a balance that results from the continual and delicate compensation by the most sensitive scales'. There must be regulatory mechanisms, on one hand, to protect the cells from any inopportune variation, and on the other, to coordinate individual activities in the common interest. The parts have to function in harmony with the whole. And Claude Bernard turned again to the example of a factory in order to describe the phenomena of regulation in living beings.

The organism represents . . . what takes place in a gun-factory, for instance, where each workman makes one part independently of his fellow-worker who is making another part, without knowing what the finished object will be. Afterwards there seems to be a fitter who puts the parts together in harmony.[9]

In the middle of the nineteenth century, the nervous system represented the 'great harmonizer of functions' in the adult animal, regulating not only the heart-beats, respiration, oxygen content and body temperature, but also concentration of salt and water, the chemical activity of the liver, the secretion of saliva, sweat and so on. At the beginning of the twentieth century were added other regulatory mechanisms, of a chemical nature: hormones. With the work of Cannon, this coordination and constancy of the internal medium became 'homeostasis'. The concept of regulation is one of those on which modern biology is founded in its most varied aspects. Thanks to this concept, biology was able for once to provide a model for physics: systems observed in living beings were used by Wiener as a basis for the development of cybernetics.

Towards the middle of the nineteenth century, it thus became possible, thanks to the methods of physiology, to analyse many different aspects of the function of organisms. Heredity and reproduction excepted, however. For, according to Claude Bernard, 'heredity is not an element that we hold in our power or that we can master as we have mastered vital properties'. The theory of evolution had turned reproduction into the mechanism responsible for both maintaining and varying structures. The cell theory had placed this mechanism inside the cell, more particularly in the egg. But heredity and reproduction offered little foothold for physiological experiments. It was certainly possible to interfere with the embryo, to damage certain cells or tissues and to impede development. But usually, the main result was to kill the embryo or rudely disturb its arrangement. Never was morphogenesis re-oriented in a direction inconsistent with the nature of the egg. A rabbit's egg could be subjected to any imaginable treatment; it could be destroyed or aborted, but 'a dog or some other mammal can never be produced from it'.[11] The recipes that had proved successful in physiology were not applicable to heredity. The study of form was no longer within the scope of chemistry; it was no longer subject to its laws. It was beyond the reach of experiment.

Able to 'contemplate' but not analyse heredity, biology was reduced to describing the formation of like by like through the use of images drawn from the preceding century. Seeds were replaced by ovules and spermatozoa, organic molecules by cells, the formation of the embryo from which the adult emerged was the result of cell division and differentiation. But for an egg to reproduce the parent organism, for an identical being to be fashioned through a succession of generations, a system of memory was needed to guide the cells.

'The egg is a becoming, it is a kind of organic formula summing up the being from which it has come and whose evolutionary memory it keeps, as it were.'[12] Evolution at that time meant the series of transformations occurring in the course of embryonic development. Memory meant a 'hereditary force', an 'anterior state' of the egg itself, a 'primitive impulse' that necessarily led the cycle from hen to egg and egg to hen. As to the nature of this memory, it was scarcely different from what had been proposed in the eighteenth century. Like Maupertuis before them, Darwin and Haeckel made memory a property of the particles forming the organism. For Darwin, each cell in the body of the begetter sent a germ or 'gemmule' to the reproductive cells, a sort of ambassador to represent and somehow constitute the cell in the following generation. For Haeckel, the cells contained particles or 'plastids' that performed specific movements; endowed with memory, throughout successive generati ns they maintained the particular movement that expressed their activity. In contrast, Claude Bernard, like Buffon, placed memory, n t in the particles constituting the organism, but in a special system guiding the multiplication and differentiation of the cells and the gradual formation of the organism. The egg thus contained a 'design' transmitted by 'organic tradition' from one being to another. The formation of the organism conformed to a 'plan' executed acc rding to very strict 'instructions'. This plan directed not only the development of the embryo, but also the functioning, structure and properties of the future adult down to the last detail, since in man certain diseases reappeared from father to son. From the egg onwards, everything was coordinated, everything was foreseen, not only for the development of the new being, but for its maintenance throughout its entire life. In Claude Bernard's words, 'Every act of a living organism has its purpose within the compass of the organism.'[13]

Experimental physiology, therefore, remained powerless to study reproduction. With no means to convert hypotheses into experiments, no techniques adapted to the situation, no material suitable for its own methods, it could not grapple with heredity. However appropriate for breeding and agriculture the methods of hybridization could appear, they did not seem suitable for investigating reproduction. Physiology, in fact, is concerned with the individual. One organism at a time is examined to observe its properties, behaviour and reactions in different circumstances. It was observation not of individuals, but of populations linked by bonds of kinship, that opened up the path to the experimental study of heredity. Darwin had already used this approach to explain variation as the result of the statistical fluctuations that are bound to occur within large populations. By following the behaviour of a few characters in successive generations of large populations, Mendel was able to demonstrate the phenomena of heredity, to measure them and draw up their laws. But it was primarily physics that, in order to handle the enormous masses of molecules that constitute bodies, made chance the law of the universe.

Statistical Analysis

In its early days, biology had deliberately divorced itself from physics. After the middle of the nineteenth century, the links were re-established through thermodynamics. First, because with the concepts of energy and its conservation, one of the peculiarities of the living world disappeared. Secondly, because in seeking to relate the properties of bodies to their internal structure, statistical mechanics changed the way of looking at objects, organisms and even the events of daily life. During the first part of the nineteenth century, mechanical phenomena were always analysed in terms of space, time, force and mass. 'Force' was introduced as a cause of motion, pre-existing and independent of it. For Carnot, there were two ways of considering the principles of mechanics. 'The first is to consider mechanics as the theory of forces, that is to say, of causes which impart motion. The second is to consider the theory of forces as the theory of motions themselves.' While matter was uniform, forces were continually increasing in number, nature and variety. Those phenomena in which heat was manifested became more and more closely associated with the motion of bodies, for example in the study of gases. Thus a new field of physics was formed, in which, by measuring changes in heat, physicists tried to analyse the relations between the properties of bodies without knowing their detailed structure. The forces acting in fields as different as motion, electricity, magnetism, heat, light or chemical reactions found a common denominator in the concept of energy. Energy is any form of work and it is anything that produces work or is produced by work. In the absolute, energy is indestructible, like matter, but it can undergo transformations which make it appear in various forms. The principle of conservation of energy turns each change in nature into a conversion of energy. It considers the different forms of energy as independent and of equal value. Each form of energy has a corresponding intensity factor – height for gravitation, temperature for heat, potential difference for electricity. Variations in a system result from differences between these factors.

The conservation of energy did not, however, explain the contradiction between certain events observed in physics. The phenomena of mechanics or electrodynamics are reversible; they can take place in either direction, and in the equations of mechanics the sign (+ or —) of the time-variable plays no part. Thermal or chemical phenomena, however, are irreversible: they always proceed in one direction; for example, heat cannot flow from cold to hot. In a 'closed system', the quantity of energy remains constant. But quantity of energy alone does not characterize the system; there is also quality of energy. The more usable the energy is and the better it can be converted into work, the higher its quality. There are thus 'nobler' forms of energy, as in mechanics, and other 'baser' forms, like heat. But in a system left to itself, the quality of the energy tends naturally to be degraded, not improved. Hence the one-way direction imposed on certain phenomena. When heat flows from hot to cold, it is because the energy decreases in quality without changing in quantity. Like a ball on a staircase, it has to go down and stops only at the bottom. This equilibrium state is called the level of maximum 'entropy' by physicists. Entropy is not a vague concept. It is a measurable physical quantity like the temperature of a body, the specific heat of a substance or the length of an object. It makes it possible to describe exactly the variations of state that a body or system may undergo. If a body receives heat, its entropy increases; if it loses heat, its entropy decreases. The second law of thermodynamics, by which the physical phenomena of the universe are governed, states that in an isolated system energy tends to be degraded and therefore the entropy tends to increase: in the end, motion ceases, differences of electrical or chemical potential are annulled and temperature becomes uniform. Without an external supply of energy, every physical system decays and moves towards a state of complete inertia.

The concepts of thermodynamics completely upset the notion of a rigid separation between beings and things, between the chemitsry of the living and laboratory chemistry. With the concept of energy and that of conservation, which united the different forms of work, all the activities of an organism could be derived from its metabolism. Everything that a living being could accomplish in terms of movement, electricity, light or noise became the result of the conversion of chemical energy released by the combustion of foodstuffs. There were, then, two generalizations that brought biology nearer to physics and chemistry: the same elements compose living beings and inanimate matter; the conservation of energy applies equally to events in the living and in the inanimate worlds. Those who, like Helmholtz, grasped the universality of these principles, drew a simple conclusion: there is no difference between the phenomena occurring in living beings and in the inanimate world. At first sight, living beings, by their growth, development and ability to maintain their structures through successive generations, seem to contravene the second law of thermodynamics which causes the continual decay of the universe. But though thermodynamics imposes a general direction on a system, it does not exclude local exceptions, nor forbid a counter-movement of certain components at the expense of their neighbours. It is the system as a whole that decays, not its individual parts. Because they receive energy from their surroundings in the form of food, living beings are able to preserve their low level of entropy throughout time. They can also, without breaking the laws of thermodynamics, continually produce the large specific molecules which characterize them. The concepts of energy and its conservation played one of the roles that biology had previously attributed to vital force. At the beginning of the nineteenth century, an organism expended vital force in order to perform its work of synthesis and morphogenesis; at the end of the century, it consumed energy.

The study of large populations and the introduction of the statistical method for their analysis had still greater consequences for the way biology and other sciences considered beings and things. In the nineteenth century, the study of gases made it possible to link heat to the motion of particles, and thus to connect the properties of a body with its internal structure. A gas could be regarded as a collection of molecules in free movement. For Bernoulli, Joule and Clausius, all particles have the same speed: this allows a network of relationships to be established between certain properties of the gases, such as pressure, temperature and density. For Maxwell, on the contrary, particles cannot all have the same speed, since their motions are caused by collisions between them. A gas represents a collection of 'small, hard and perfectly elastic spheres, acting on one another only during impact'. A purely mechanical model of a gas can be made: the particles continually travel a certain distance, collide, start again, collide and start off once more. Each particle, then, has a unique velocity and motion. According to Maxwell, 'each carries its energy and its motion with it'. And the characteristics of each particle vary continually as a result of random collisions. There is therefore no question of studying in detail the ever changing behaviour of each of the billions of individual gas molecules. On the other hand, the whole population of molecules can be treated and its behaviour analysed by statistical methods. The speeds of molecules then must be distributed in accordance with the well-known probability curve that applies to phenomena as different as the heights of adults in a country, the number of puppies in a litter, or the scatter of shot from a gun. The behaviour of individuals cannot be described, but that of the population can. It is possible to think of the population as consisting of ideal molecules; their parameters are the averages of the real value. The properties of a gas can be described by the purely mechanical model of balls that collide, and even entropy can be interpreted in terms of molecular agitation. If man is unable to prevent the degradation of energy, it is because he is unable to distinguish each molecule and observe its characteristics. But it is perfectly possible to imagine a being with a better brain and finer senses, whose faculties, according to Maxwell, 'are so sharpened that he can follow every molecule in its course; such a being, whose attributes are still as essentially finite as our own, would be able to do what is at present impossible to us'.[14] This tiny being or demon has to be imagined as capable of 'seeing individual molecules', and able to move a sliding door which causes no friction, in a partition separating two compartments of a gas-filled vessel. When a rapidly moving molecule arrives from left to right, the demon opens the door; when a slow moving molecule arrives, he closes it; and conversely. The rapid molecules will then accumulate in the right-hand compartment, which will get warmer, and the slow molecules in the left-hand compartment, which will cool down. 'Without expenditure of energy' the demon will thus have converted non-utilizable energy into utilizable energy. He will have circumvented the second law of thermodynamics.

During the second half of the nineteenth century, however, the role and status of statistical analysis and of the theory of probability changed. For Maxwell, they were simply tools suitable for the analysis of a particular problem. Since it was impossible to observe each individual, the whole population had to be considered. For Boltzmann and Gibbs, on the contrary, statistical analysis and the theory of probability supplied the rules for the logic of the whole world. Large numbers are studied not so much because it is impossible to investigate the individual units, but mainly because their behaviour is of no interest at all. Even if they could be analysed in detail and subjected to a mathematical treatment, individual cases could teach no more than the population taken as a whole. What could be the advantage of knowing the distance travelled by a particular molecule? Or of learning that a given molecule in a gas-filled vessel has collided with the wall at a given time and place and under certain circumstances? Even if the detailed behaviour of each unit could be analysed, the mass of results could only be put together to give the statistical law that governs the population as a whole. There is no point in knowing which particles collide at a given moment; what is important is the average number of collisions and the probability of any given particle participating in one.

This approach is obviously completely different from all those which had preceded it, that of Darwin excepted. For Darwin, as for Boltzmann and Gibbs, the laws of nature apply not to individuals but to large populations. Although there might be irregularities in the behaviour of each unit, in the end the large numbers involved impose regularity on the whole. The analogy between the two ways of thought goes even further. First, with statistical mechanics, as with the theory of evolution, the notion of contingency became established in the very heart of nature. Since Newton, physics had been based on a rigid determinism. The behaviour of molecules, like that of all visible bodies, was considered to be strictly controlled by a system of causes, which science tried to introduce into the laws of nature. If observable phenomena were to be repeated exactly, the processes from which they came, the elementary processes, had also to be subject to fixed determinism. But in the second half of the nineteenth century, several so-called laws of nature became statistical laws. They are strictly obeyed in so far as the number of individuals involved is very large. What these laws predict cannot be formulated in terms of strict causality; it is merely probable, and can be verified only within certain well defined limits. With observable phenomena, this probability borders on certainty, because visible bodies are composed of enormous numbers of molecules. But with less important populations, deviations are not rare; they are what Boltzmann called 'statistical fluctuations'. If there is a particular mechanism that favours them, like natural selection in evolution, then the exceptions will eventually predominate.

Finally, the analogy between the theory of evolution and statistical mechanics extends to include the notion of the irreversibility of time. In evolution, the mechanism of selection makes the whole process irreversible: once certain variants have gradually been selected, once a group of organisms is pointed in a certain direction, there is no chance of the group ever returning to its former state. Natural selection can still accentuate differentiation, it can even modify the direction, but it cannot reverse the steps that have already occurred.

In physics, the second law of thermodynamics imposes a direction on phenomena; no event can go in a direction different from that observed, for that would mean a decrease in entropy. No part of the universe's substance can return to a former condition, as might be imagined in a purely mechanical system such as an imaginary clock. In neither the organic nor the inanimate physical world can the film sequences disclosing evolution be run backwards.

There remained, however, some flavour of mystery in the processes by which energy is degraded. It was as if irreversibility required some secret component common to the different mechanisms found in nature. With the development of statistical thermodynamics, the need for a hidden factor disappeared. Irreversibility expressed changes in the order and arrangement of molecules. The one-way movement became the result of a property inherent in the very structure of matter. For with Boltzmann, the second law of thermo-dynamics, that controls the march of the universe and produces on balance an increase of entropy, was ultimately only a statistical law. It became, in fact, *the* supreme statistical law. Most physical phenom-ena simply express the natural tendency of populations of molecules to pass from order to chaos. For a physicist, molecular order is a statistical value which can be measured. The heat stored in the sun, for example, is an enormous stock that can be used because it is not distributed uniformly throughout the universe, but remains con-centrated in a limited space. In time, this heat tends to disperse spontaneously and the temperature to become uniform throughout the universe. This is equivalent to an increase in disorder or entropy. Heat tends to flow from hot to cold, not because of a secret law for-bidding it to flow in the opposite direction, but only because the reverse flow – cold to hot – is less probable by many orders of magnitude: it never happens in practice, without being absolutely impossible in theory. To speak about molecules passing from a less probable to a more probable state is like speaking about the stones of a monument that an earthquake turns into a heap of ruins, or books carefully arranged on library shelves and then muddled by careless users. Statistical thermodynamics states that when a pack of cards is shuffled, they are most likely to pass from order to chaos. It does not state, however, that the converse is impossible. This can, and indeed must, occur with a large enough number of attempts. For it to occur, however, so much time is needed that such exceptions cannot upset the overall forward march of the universe. The stream of events flows towards what is most statistically probable. For Darwin, the irreversibility of evolution stemmed from the impossibility of a return to an earlier state for organisms already committed to a particular path of specialization. For Boltzmann, the thermo-dynamic irreversibility stemmed from the impossibility of a return to a previous state for the molecules of the universe, once they had passed spontaneously from order to disorder.

The whole attitude of the nineteenth century was transformed by the new outlook inspired by statistical mechanics. In the first place, statistical mechanics derived the properties of bodies from the very structure of matter. With Gibbs, statistical analysis applied not only to the behaviour of large populations, but to all 'conservative systems' with any degree of freedom. It permitted analysis of the distribution of positions and moments compatible with the energy of a given system – a distribution that included the whole system if it functioned long enough. Most events which occurred in the physical world could be treated in this way: chemical reactions, their velocities, their variations with temperature, the processes of fusion and evapor-ation, the laws of gas pressure and so on, were all phenomena based on the underlying hypothesis of changes occurring in molecular order. All became subject to statistical laws.

With statistical mechanics came the perfection of the mathematical tool that made it possible to investigate the structure and evolution of any system involving large numbers. Many objects, events and even properties which had hitherto evaded analysis could now be dealt with in so far as they could be enumerated and classified in a discontinuous system. This kind of statistical analysis is, in fact, based entirely on the distribution of discrete units. Whether this discontinuity exists naturally, as in populations of units, or is intro-duced by the methods of measurement which require a choice be-tween two limiting values, it is a necessary condition for this kind of analysis. For discontinuous things can be counted by means of the oldest and simplest mathematical concept – whole numbers. Ability to count whole numbers is the art of applying statistical method. The greater the number of cases observed, the more reproducible are the results. But statistical method is so sure, it functions with such precision, that conditions can be adjusted so that only a limited number of observations is required. With Maxwell, statistical method was applied to physical phenomena mainly as a tool. After Boltzmann and Gibbs, the method was gradually extended to a variety of fields, even to those where it seemed at first difficult, if not impossible, to introduce the necessary discontinuity. It became possible to derive practical laws from phenomena of which the determinism was not known. Instead of seeking the causes of isolated events, it became possible to observe a large number of events of the same class, sort them, collect the results and then calculate the mean by empirical rules. Future events in the same class could then be predicted, not with certainty, but with a probability that very often amounted to certainty. These predictions were valid only for the totality of events, excluding details and exceptions. In fact, one of the characteristics of the statistical method is that it deliberately and systematically ignores details. It matters little whether every circumstance is minutely described, or whether every possible detail of information is obtained; this is not the aim. The aim is to obtain a law which transcends individual cases.

Finally, statistical thermodynamics completely transformed the way of looking at nature, mainly because it brought together and gave the same status of related and measurable quantities to order and chance – two concepts which until then had been incompatible. The whole range of forces, impulses, changes and potentials, which in spite of everything still retained some slightly mysterious and arbitrary flavour, were relegated to the rank of auxiliary factors. They simply represented different aspects of a more profound and universal mechanism which emerged as the general law of the universe: the natural tendency of things to pass from order to disorder, through the effect of calculable chance. This law does not aim to give a causal explanation of events; it does not say *why* they occur, but *how*. From then on, the very notion of causality lost something of its significance, and even of its interest. As a result, much of the mystery which still pervaded the picture of nature in the first half of the nineteenth century vanished. Many totally different and unexplained phenomena often exhibit common characteristics, because in one way or another they are based on a common mechanism. This is true not only of the phenomena investigated by physics. At the end of the nineteenth and the beginning of the twentieth century, it also applied in astronomy, geology, biology, meteorology, geography, history, economics, politics, industry and commerce; in fact, in a wide field of human activities and even in the details of daily life.

It is no exaggeration to say that the way we now regard nature has to a large extent been fashioned by statistical thermodynamics, which has transformed both the objectives and the outlook of science. It has brought about the change of attitude from which was born, at the beginning of the twentieth century, the world of modern physics: a world of relativity and uncertainty subject to quantum laws and information theory, a world in which matter and energy are merely two aspects of the same thing. It was through statistical thermodynamics that new sciences emerged – such as physical chemistry which bases the chemical properties of bodies on their physical structure. Again, thanks to statistical thermodynamics, experimentation could be extended to the most varied fields of biology; first because the chemical reactions that occur in living beings obey the laws controlling matter in general; also and above all because statistical analysis transformed biology into a quantitative science. At the end of the nineteenth century, the study of living beings was no longer a science of order, but one of measurement as well.

The Birth of Genetics

The attitude of Darwin, Boltzmann and Gibbs did not express an idea confined to very few people, but rather a tendency which became prevalent after the middle of the nineteenth century, as shown by the work of Mendel. Over the centuries, observations about heredity had accumulated. Nonetheless, heredity was not, strictly speaking, a real object of investigation. The nineteenth century was no longer bent on verifying the existence of fabulous monsters; it had abandoned attempts to hybridize different species. Nevertheless, crosses were still undertaken between varieties that differed with respect to a large range of characters. Heredity was principally the concern of horticulturists and animal breeders. In fact, throughout the entire nineteenth century, economic necessity forced increased yields of crops and herds, as well as the development of varieties adapted to special local conditions. Production had to be stepped up, not only by increasing the yields of animals and plants per acre, but also by improving quality. It was in orchards and pastures, beehives and poultry-yards that the practical experiments were carried out. After each cross, it was sufficient to scrutinize the progeny closely in terms of all their recognizable traits, so as to make their description as complete as possible without omitting any detail. Most characters examined could not be precisely distinguished, but shaded into one another through an almost infinite series of intermediate types. In fact, the success of the operation was judged by the degree to which the characters of the parents blended in the hybrid. Thus characters were seen to be reassorted in successive generations. Some disappeared for a time, only to reappear later. Naudin, for instance, contrasted the uniformity of first-generation hybrids with the 'extreme confusion of forms' in the second generation: some resembled the father, others the mother, as if the hybrids were 'living mosaics'[15] whose components were not perceptible to the naked eye. Gärtner observed wide heterogeneity in the offspring of hybrids: some produced pure descendants, others produced mixtures. The principal attribute of heredity seemed to be its complexity.

Mendel represented the meeting point of the two currents that led to the establishment of a science of heredity: practical knowledge of horticulture and theoretical knowledge of biology. The son of a farmer, he became interested in evolution. Throughout his youth, he saw his father planting, crossing and grafting. All his life, he wondered how species were formed. In the garden of the monastery where he lived, he obtained permission to cultivate a few plants. What chiefly fascinated Mendel, however, was the nature of heredity which the vigour of grafts showed to be stronger than environment, that is the stock on which they had been grafted. He too began to produce hybrids, not to improve the yields, but to follow the behaviour of characters from generation to generation. Mendel's attitude was entirely different from that of all his predecessors. 'Among all the numerous experiments made,' he wrote,

not one has been carried out to such an extent and in such a way as to make it possible to determine the number of different forms under which the offspring of hybrids appear, or to arrange those forms with certainty according to their separate generations, or definitively to ascertain their statistical relations.[16]

There were three entirely novel elements in Mendel's approach: the way of envisaging experiments and choosing appropriate material; the introduction of discontinuity and the use of large populations, which meant that results could be expressed numerically and treated mathematically; the use of a simple symbolism, which permitted a continuous interchange between experiment and theory.

First, Mendel selected his material with great care. He tried several plants before deciding on the Pea. Then he used varieties whose purity was guaranteed by several years' culture under stringent conditions. The variants to be hybridized had to differ from one another, not as a whole, but in a limited number of traits. According to Mendel, characters that 'do not permit of a sharp and certain separation' must be eliminated, 'since difference is of a "more or less" nature which is often difficult to define'.[17] Only those traits must be retained which can be distinguished without ambiguity, such as the form and colour of the seeds and pods, the arrangement of the flowers on the stem, etc. To avoid from the start insurmountable complexity in the analysis of hybrids, details ought to be ignored and the study confined to a very small number of characters: first one, then two, then three, taking care each time to distinguish all the possible combinations in the offspring. To exhaust all possible combinations, two conditions have to be respected: firstly, experiments have to be made on a scale sufficient that individuals can be ignored and only populations taken into consideration; secondly, the behaviour of the characters has to be followed, not only in the first generation, but in a long series of successive generations.

By its very nature, this type of experimentation led to an entirely novel way of expressing the results. Because of the discontinuity deliberately introduced into the discrimination of characters, it was sufficient in each generation to count the individuals of each of the possible classes. Each class was thus expressed by a whole number, and the wider the scope of the experiment, the larger this number. These numbers could be treated statistically and the relation between them established. The numbers generally appeared in simple ratios. Thus in crosses between varieties differing in a single character, first-generation hybrids resemble only one of the parents, never the other: the character of the latter is said to be 'recessive' compared with that of the former or 'dominant' parent. In the next generation produced by these hybrids, the two forms, recessive and dominant, appear in a ratio of about $1:3$. According to whether the carriers of the dominant character produce offspring identical to themselves or continue instead to produce recessive forms, they can be divided into two further classes in the ratio of $1:2$. When the varieties used differ, not by one, but by two characters, hybrids are again all identical. In the next generation, the offspring of the hybrids fall into four classes in ratios of about $1:3:3:9$. Three of these classes can again be subdivided each into two during following generations. And the number of classes increases with the number of characters involved. In Mendel's words,

The offspring of the hybrids in which several essentially different characters are combined exhibit the terms of a series of combinations, in which the developmental series for each pair of differentiating characters are united . . . If n represents the number of the differentiating characters in the two original stocks, 3^n gives the number of terms of the combination series, 4^n the number of individuals which belong to the series, and 2^n the number of unions which remain constant.[18]

In other words, the different characters are transmitted independently. In sufficiently large populations, the distributions of characters can be predicted.

Finally, the simple binary choice between the two forms of a character makes a simple symbolic representation possible. According to Mendel,

If A be taken as denoting one of the two constant characters, for instance the dominant, a the recessive, and Aa the hybrid form in which both are conjoined, the expression $A + 2\ Aa + a$ shows the terms in the series for the progeny of the hybrids of two differentiating characters.[19]

The symbolic interpretation of the results in some way becomes the hinge between theory and experiment. It permits hypotheses to be formulated easily from the observed distributions; and it leads directly to predictions which can then be experimentally tested. In this way, from the observed relations between the combinations of characters it is possible to draw conclusions about the formation and constitution of the pollen and egg cells. A line of descendants remains pure and constant only if the organisms come from pollen and egg cells bearing the same characters, for example A. There is no reason to think that another mechanism operates in the formation of hybrids, for example Aa. Since both forms of the same character A and a are produced in the same hybrid plant, or even in the same flower, it must be concluded that in the ovaries of the Aa hybrid, equal numbers of egg cells of types A and a are formed, and in the anthers equal numbers of pollen grains of the same two types. More generally, when several characters are involved, as many kinds of pollen and egg cells must be formed in the hybrid as there will be combinations in the offspring. This influence is confirmed by experiment. According to Mendel, in a hybrid Aa

it remains purely a matter of chance which of the two sorts of pollen will become united with each separate egg cell. According, however, to the law of probability, it will always happen, on the average of many cases, that each pollen form, A and a, will unite equally often with each egg cell form, A and a; consequently one of the two pollen cells A in the fertilization will meet with an egg cell A, and the other with an egg cell a, and so likewise one pollen cell a will unite with an egg cell A, and the other with egg cell a.[20]

The results of these crosses can even be described by a simple graph. But to represent the character of an individual, a simple symbol is no longer enough; two are needed, which Mendel brought together in the form of a fraction. In the offspring of hybrid A/a, four combinations are formed: A/A, A/a, a/A, a/a. Only the first and last, which correspond to the characters of the parents, are pure. Since form A is dominant over form a, the first three have the same *observable* character, although a different structure, revealed in their offspring; hence the $1:3$ ratio between the recessive and dominant forms in the second generation. These values represent only the average result of numerous experiments on self-fertilization of hybrids. In individual flowers or plants, the values of the series often differ from the average. According to Mendel,

The separate values must necessarily be subject to fluctuations ... The true ratios of the numbers can only be ascertained by an average deduced from the sum of as many single values as possible; the greater the number, the more are merely chance effects eliminated.[21]

With Mendel, biological phenomena suddenly acquired the rigour of mathematics. A whole internal logic was imposed on heredity by methodology, statistical treatment and symbolic representation. Apart from the episode of preformation, ideas about the mechanism of heredity had scarcely changed in two thousand years. The theory of evolution demanded a process able to reproduce parental traits in the offspring, as well as to vary them slightly. What in Mendel's time Darwin envisaged as 'pangenesis' strongly resembled what had already been imagined by Hippocrates and Aristotle, and later by Maupertuis and Buffon. According to the theory of pangenesis, each fragment of the body, each cell, produced a little germ of itself, or 'gemmule', that was sent to the germ cells and commissioned to reproduce the same fragment in the next generation. This theory had the advantage of allowing for the possibility of spontaneous variations uninfluenced by external factors, as well as for the insertion of acquired characters into heredity. Darwin, no more than Maupertuis and Buffon, did not distinguish between what constitutes the body of the parents, their seed and the body of the child. Delegated by the parents, the same components passed into the seed to form the child. Heredity, in consequence, could be located only within organization itself, in that secondary structure to which all perceptible structures and functions of a living being were attributed. Mendel looked at heredity in a completely different way, in terms of phenomena that could be analysed with precision. Regular segregations, dominance of characters, persistence of the hybrid state, none of these are in accord with pangenesis. To represent a recognizable feature in an individual, two symbols are needed. One symbol, therefore, cannot correspond either to an observable character or to its delegate, the gemmule. Hence the necessity of distinguishing between what is seen, the character, and something else underlying the character; between what twentieth-century genetics call phenotype and genotype. Genotype determines phenotype, but is only partly expressed in it. Observable characters simply bear witness to the hidden presence of particles or units which Mendel called 'factors'. These are independent of each other, and each determines an observable character. A plant has two samples of each factor, one coming from each parent, either through pollen or the egg cell. What is transmitted by heredity, therefore, is neither a complete representation of the individual nor a series of ambassadors from all parts of the parental bodies, subsequently rearranged in the offspring like stones in a mosaic. It is a collection of discrete units, each controlling one character. Each unit can exist in different states that determine the different forms of the corresponding character. Since each organism receives a complete set of units from each parent, the units are re-assembled haphazardly in the course of generations. The organization which anatomists, histologists and physiologists were studying, that second-order structure to which all the forms and properties of a living being were ascribed, was no longer enough to explain heredity. A higher-order structure has to exist, still more hidden, more deeply buried in the body. It is in a third-order structure that the memory of heredity is located.

Accordingly, the very attitude that led Boltzmann to link the properties of bodies to their internal structure, in order to derive the law that determines the evolution of matter, provided Mendel with a means of studying heredity and learning its laws. In both cases, discontinuous elements are involved. In both cases, the behaviour of a single element subject to chance cannot be foreseen. And in both cases, statistical treatment of large populations makes it possible to extract order from chance. For hereditary factors and gas molecules alike, the behaviour of each individual unit is unimportant. The combination of characters in a particular plant was of no more interest to Mendel than the path of a particular molecule was to Boltzmann. This was the price which had to be paid before heredity could be subjected to analysis. The humours, obscure forces and mysterious purposes that ever since antiquity had seemed to fashion the characters of living beings could be replaced by matter, particles and laws. The entire representation of living organisms was thus transformed. Logically, the whole practice of biology should also have been transformed. In fact, nothing of the sort occurred. Mendel's case represents a good example of the impossibility of tracing a linear history of ideas, of finding the succession of stages that logic would have deliberately followed. For although it was in accord with the physics of his time, Mendel's work did not have the slightest influence on the way his contemporaries studied biology. It was the twentieth century that made Mendel the creator of genetics and transformed his first paper into the birth certificate of that science. Until the turn of the century, his work remained unknown or neglected. Not that Father Gregor Mendel was unknown to scientists of his day. He was perhaps an amateur, but he was in contact with many of the most famous biologists of his time. He kept up a long correspondence with several of them, describing his experiments in detail; but he did not capture their attention. When Mendel read his first communication to the local society of natural sciences one evening in February 1865, there were about forty people in the Realschule in Brno. They included naturalists, astronomers, physicists and chemists – in other words, a knowledgeable audience. Mendel spoke for an hour about the hybridization of peas. His audience felt kindly towards the lecturer himself. Although surprised that arithmetic and calculation of probabilities entered into the question of heredity, they listened patiently and applauded politely. When Mendel had finished his report, everyone went home without expressing the slightest curiosity. Mendel wrote to Nägeli: 'I encountered, as was to be expected, divided opinion; however, as far as I know, no one undertook to repeat the experiments.'[22] When Mendel died a few years later, he was honoured for his social functions but ignored as a scientist. At the beginning of the twentieth century, when his work was 'rediscovered', the pages of the review containing Mendel's paper were often found uncut.

How can it be said, then, that the human mind is only waiting to seize new ideas and exploit them? How can we possibly consider the development of sciences to be guided by the finality of logic alone? Logic can manoeuvre only within the area defined by the outlook of the epoch; it can analyse only those objects which are deemed worthy of investigation. Genetics could not arise until, at the end of the nineteenth century, the study of the cell had undergone a radical transformation. The analysis of its structure had to be refined, revealing the chromosomes and their movements, as regular as those of a ballet; also the analysis of its role, which replaced the mechanism of pangenesis by the 'germ cell', a cell line reserved solely for the purpose of reproduction, protected from the vicissitudes that affect the body.

The Dance of the Chromosomes

After the middle of the nineteenth century, the cell became a focus of biological research. It was no longer merely the unit of structure of all living organisms, the final point of anatomical analysis. It had become the place where all the activities of the organism were conjoined, the 'seat of life', in the words of Virchow. In the cell, metabolic reactions take place and the characteristic molecules of living beings are fashioned. Through cell differentiation, organs are formed and the body of the adult is constructed. By cell division, organization is perpetuated. There is no cell that does not arise from another cell. Reproduction is accomplished by an 'outgrowth of the individual', in Haeckel's expression. As clearly illustrated by unicellular organisms that multiply by fission, the phenomena of heredity are really an expression of the phenomena of growth. Each little organism grows until it can divide into two parts, identical not only in size but also in form and structure. It is easy to understand why the offspring resembles its parent: the offspring is a fragment of the parent. The situation is not different in multicellular organisms, produced through the multiplication of a single initial cell, the egg. The body of such an organism may be compared to a colony of cells, in which the division of labour calls for specialization of the units. This means that certain cells are able to perform only the functions indispensable to respiration, for example, and others those of reproduction, locomotion or digestion. Whether in a unicellular organism or a complex one, heredity always results from cellular continuity. The cell itself is, therefore, the site both of the chemical reactions which give the organism its specificity and of the system which endows it with the capacity to produce its like. The germ cells contain the rough outline of the unborn organism, not in effigy, but in potential form. In these drops of albuminous substances, the specialized pattern of all the cells of the future organism are already enclosed. The focal point of investigation therefore became the functioning and division of the cell.

It was not a question of constructing a zoology of cells; of defining the position, relations and properties of all the units composing an organism; of learning their exact affiliation; or of constructing their detailed map in the body. In fact, there was little chance of ever being able to unravel the network of cells and fibres in a complex organism. But although the cells in an animal have different shapes, locations and tasks, they are nevertheless built on the same model. In spite of their diversity, they exhibit the same structure. Whatever its nature and origin, a cell is always a semi-liquid corpuscle composed of albuminous material, protoplasm. It always contains a more or less small, round nucleus, also made of albuminous material. It is often surrounded by a membrane and sometimes filled with particles. The organization of the cell is dominated by the presence of two major components. 'Nucleus and protoplasm,' said Haeckel, 'internal cellular nucleus and external cellular substance are the only two essential parts of every real cell. All the rest is secondary and accessory.'[23] It is, therefore, necessary to assign roles to each of the two constituents and to define their composition and function in order to identify what it is that the cell transmits to its descendants, so that they are formed in its image.

Cytology, the science that attempts to chart cellular space, unites very diverse interests: physiology, embryonic development, heredity and evolution, just as much as morphology. Its unity results from its method, language and material. At the end of the nineteenth century, the light microscope had attained the maximum resolution permitted by physics. By the use of substances that selectively stain certain cellular structures, however, cytologists increased their means of discrimination and identification. In this way, they even obtained an insight into the chemical composition of cell constituents: the nucleus, for example, is easily stained by certain basic substances. Gradually the landscape revealed by the microscope was thrown into relief. Its detailed description required a new language. At the end of the century, a whole new vocabulary was fashioned by juxtaposition, or even hybridization, of Greek and Latin roots. The biologist's discourse soon became incomprehensible to the layman. The ease of staining the nucleus, for instance, was indicated by the radical 'chrome'. Hence 'chromatin', as Flemming called the substance contained in the nucleus; 'chromosomes', as Weldeyer called the filaments visible inside; 'chromomeres', as Balbiani and Van Beneden called the bands that make cross-bands on the chromosomes; chromatids, chromidia, chromidiogamy, chromioles, chromocentres, chromonemes, chromoplasts, chromospires, and so on. The precision of knowledge was reflected in the precision of names. Lastly, cytology was characterized by the material it used. Since it was seeking to analyse, not the characteristics of special cellular types, but the attributes common to all cells, it had complete freedom of choice. It was therefore able to concentrate on a few organisms, particularly advantageous for observation and experimentation. On these privileged organisms was focused the attention of specialists in many different countries, of biologists from different disciplines and with different interests. Two kinds of material appeared particularly suitable. First, unicellular organisms, Protista, whose reproductive cycle is similar to that of multicellular organisms: instead of uniting to form a single body, however, protist cells remain separate and live independently. As Richard

Hertwig said, 'In Protozoa there is only one kind of reproduction, namely, cell-division.'[24] Physiologically, a protozoan constitutes an individual organism in the same way as a metazoan; by its morphology and method of formation, however, it can be compared just as much to a germ cell as to any other cell of a metazoan. Protozoa thus provide a simple material, particularly suitable for the study of cell-division, which they reveal in its purest and most stripped-down form. Secondly, among the almost infinite variety of multicellular organisms, some lend themselves particularly well to observations on the cell nucleus, on germ cells and on embryonic development. It is from these two kinds of privileged organisms that a material must be selected in terms of the specific goal. If cell-division or the morphology of the nucleus or its mode of formation is to be studied, then the organism of choice is the *Ascaris*, a parasitic worm of the horse, whose qualities were revealed by Van Beneden and Boveri. In the words of Boveri,

Ascaris forms an unsurpassable material. The eggs can be stored for some months, dry, in the cold, without alteration. When one has time for work on them, this can be done at room temperature, where they continue to develop slowly. If one wishes to accelerate development temporarily, one brings the egg into an incubator. If one must interrupt work, one puts them back in the cold, and, on returning, one finds them in the same condition in which they were left.[25]

And above all, the nucleus of Ascaris is particularly simple; the number of chromosomes is small, generally four, and even two in a certain type; it is easy to recognize them, to observe their shape and behaviour, to watch them splitting into two and arranging themselves along a sort of spindle that attracts them to two opposite poles when the cell divides; in short, Ascaris is the ideal organism for investigating the mechanism by which one cell produces two similar cells. But, if one wishes to study germ cells, fertilization and embryonic development rather than cell-division, then the organism to be used is the frog, or even the sea-urchin, whose advantages were revealed by O. Hertwig and Boveri: the egg is transparent and easy to observe; the spermatozoon is small, with a dense readily visible nucleus. If an egg and some sperm are placed in a dish of sea-water, one can see the spermatozoa adhering to the egg. 'But only one sperm reaches this goal, the one that first touches the naked egg surface,' said Boveri.[26] It is possible to follow the path of the male nucleus that fuses with the female nucleus, and to watch the successive divisions that occur in the egg in strict temporal and spatial order – in other words, to observe in detail the kind of a miracle by which fragments detached from two individuals, one male and one female, interpenetrate and give birth to a new and identical organism.

With the sea-urchin egg, the study of the cell and of embryonic development ceased to be purely observational, and became experimental. It proved possible, in fact, to influence the germ cells or the developing egg, and even to modify the chemical and physical conditions of artificial fertilization. By shaking unfertilized eggs vigorously, the Hertwig brothers were able to break them into pieces that could still be fertilized by the sperm of the same species. By treating eggs with certain compounds, Boveri succeeded in fertilizing each of them with several spermatozoa; and when he shook these eggs, he observed abnormal distribution of chromosomes in the dividing cells. By increasing the salt concentration in sea-water or exposing eggs to various chemical or physical treatments, Loeb induced artificial parthenogenesis. By isolating one cell of a fertilized egg which had begun to segment, Driesch obtained development of a complete organism – small, perhaps, but perfect. In the twentieth century, the technical virtuosity of embryologists continually increased. They were able to operate on particular cells in an egg, to destroy them at will, to inject certain substances or even extracts of other embryos, to remove the nucleus of an egg and replace it by another. The effect was measured by the lesions that developed in the embryo, by the stage it reached in its development and by the monsters that appeared. Thus even the formation of the embryo had become open to experimental analysis.

The first task of cytology in the nineteenth century was to distinguish the part played in cell functions by its two chief components, nucleus and cytoplasm. Little by little, the nucleus came to the forefront; and within the nucleus, it was to the chromosomes that the chief role was attributed. The constancy of their numbers and shapes, the accuracy of their movements, the precision of their segregation in the products of cell-division, all combined to give them an exceptional position. Chromosomes can be seen to thicken, then to become thinner again, to disappear and reappear later with the same form as before. They can be seen to split lengthwise into two identical products, each drawn towards one pole as if by a 'magnetic centre', in Van Beneden's words. Chromosomes show continuity through the cell cycle. They have individuality and a characteristic structure at certain stages. They are 'organized elements existing autonomously in the cell',[27] as Boveri said. Above all, they have the extraordinary power, through dividing and reassembling at each pole, to form two nuclei identical with the original one. It is possible to distinguish chromosomes from one another, to follow their evolution, to count them. They go in pairs: two groups of two in Ascaris, except for the germ cells, the ovule and the spermatozoon, where each nucleus contains only two chromosomes. But the fusion of these nuclei at the time of fertilization produces a complete set of chromosomes, half from the father, half from the mother. Any inaccuracy in the number of chromosomes so reassembled, any excess, any deficiency throws embryonic development into disorder. For Boveri, 'normal development is dependent upon a particular combination of chromosomes; and this can only mean that the individual chromosomes must possess different qualities'.[28] It was, therefore, a structure with exceptional features that became apparent within the nucleus, a structure that had the property unique in the cell, of exact duplication.

At the same time, the duality first observed in the cell extended to the whole organism. Until then, no distinction was made between the cellular elements concerned with structure and those concerned with reproduction. The child represented an outgrowth of the parents, from which each part sent through the reproductive cells a kind of germ, destined to re-form precisely that part in the next generation. Thus a single particle had to exist first in an organ in the parent, then in a reproductive cell and finally in the same organ in the offspring. According to Huxley, 'It is conceivable, and indeed probable, that every part of the adult contains molecules derived from the male and from the female parent; and that, regarded as a mass of molecules, the entire organism may be compared to a web of which the warp is derived from the female and the woof from the male.'[29] But with what Nägeli called 'trophoplasm' and 'idioplasm', there appeared a duality in the organism as a whole: the trophoplasm, which formed the major part of the body, was responsible for nutrition and growth; the idioplasm, in contrast, represented only a small component in volume, but played an essential part in reproduction and development; it was the substratum of heredity. Contained within the egg, it directed the evolution and development; it spread throughout the organism, forming a kind of master-network. If a hen's egg was different from a frog's egg, it was because it contained a different idioplasm. The species was contained in the egg, just as much as in the adult organism. Idioplasm was a very complex substance consisting of an enormous number of particles or 'micellae'. According to Nägeli's calculations, a thousandth of a cubic millimetre could contain up to four hundred million micellae. It was the way micellae were distributed in the idioplasm that ensured its specificity. The reproduction of forms through successive generations was therefore no longer achieved by representatives from all parts of the body that reunited in the egg, but by a special substance directing development. Of all the biologists of his time, Nägeli was the best placed to interpret Mendel's work. It was to Nägeli, in particular, that Mendel communicated the results of his experiments in a series of letters; but without the slightest effect.

With Weismann, the distinction between these two kinds of constituents sharpened. Furthermore, it took a different form: it concerned no longer substances spread throughout the body, but cells themselves. Reproduction involved cells of a particular type, the germ cells, that differed from those constituting the body, the somatic cells, both by their function, their structure and even their role in evolution. According to Weismann, germ cells contain a substance, 'that by its physical and chemical properties and by its molecular nature is able to become a new individual of the same species'.[30] It is the quality of this substance that decides whether the unborn organism shall become a lizard or a man, whether it will be large or small, whether it will resemble its father or mother. Reproduction is based entirely on the nature and properties of germ cells. They 'are not important for the life of the individual, yet they alone preserve the species'.[31]

The proposition that a child is simply a kind of bud of the parents was, so to speak, turned on its head. According to Weismann, although germ cells can produce both types of cell, somatic cells can produce only somatic cells. Germ cells cannot, therefore, be considered a product of the organism. In the successive generations of animals, they behave like a line of unicellular organisms that reproduce by fission. From this germ line the somatic cells are differentiated. The bodies of animals are, as it were, grafted laterally on the line. Thus, as Weismann said,

The reproduction of multicellular organisms follows the same course as in unicellular organisms: a continuous division of cell – the only difference being that in complex organisms the germ cell does not form the whole individual, which is surrounded . . . by millions of somatic cells that form the outer unity of the individual.[32]

Since germ cells reproduce by fission like protozoa, they always contain the same hereditary substance. The organisms they produce must therefore necessarily be identical. The germ line forms the skeleton of the species, on which the individuals are attached like excrescences. It is no longer the hen that produces the egg. As expressed by Butler's witticism, the egg has found in the hen a convenient way of making another egg.

A further conclusion could be drawn from the roles assigned to germ cells and somatic cells. If germ cells derived directly from those of the preceding generation, if they were not produced by the body of the parent, then at the same time they were sheltered from external events. Whatever misfortune an organism might suffer, its germ cells, and therefore its descendants, were beyond reach. How, then, could characters acquired by a living organism be transmitted by heredity? According to Weismann, 'All changes due to outside influences are temporary and disappear with the individual.'[33] They are transient episodes that affect particular organisms, but not the species. The individuals forming the species have no influence on its framework. Protected from mishap, the germ cells go on reproducing identical cells. The organism can therefore acquire no character to which it is not predisposed by heredity. The whole future of an individual, its forms and properties, are already determined in the egg. Although a margin of action is still left for external conditions, it is 'restricted to a small mobile area round a fixed point that is formed by heredity'.[34] Constant in nature within the species, germ cells vary from one species to another. What is changed to produce new structures is not the individual itself, but the 'hereditary arrangements' contained in the germ cells. 'Natural selection appears to operate only on the qualities of the adult organism,' said Weismann, 'but in truth it works on predispositions that lie hidden in the germ cell.'

The way of looking at heredity was thus radically transformed. Until then, the possibility of inheriting acquired characters had never been seriously questioned. The writings of antiquity, whether Egyptian, Hebrew or Greek, were larded with stories about the perpetuation in children of accidents that had happened to their parents. On this, Lamarck had erected a whole system, making it the mechanism that explained local transformations, the helping hand that enabled the organism to adapt itself fully to its surroundings. The inheritance of acquired characters was allied to a mass of superstitions: spontaneous generation, the fertility of crosses between species; in short, all the old myths about the creation of man, animals and the earth. More than anything else, the transmission of acquired characters has resisted experimental analysis. More than anything else, it has held back the investigation of living organisms in general, and of reproduction in particular. Even for Darwin, who made evolution rest on spontaneous fluctuations occurring in every large population, pangenesis still permitted external conditions to influence directly hereditary characters. For Weismann, however, the environment can no longer direct heredity. For him, the germ line is beyond the reach of any variation that might occur in individuals of the species. None of the supposed transmissions of acquired characters stands up to analysis. None of the organisms that are mutilated generation after generation produces mutilated descendants. Even when the tails of mice are systematically cut off at birth for five generations, hundreds of little mice continue to be born with normal tails of the same average length as their antecedents. Heredity is proof against any individual whims, any influences, desires or incidents. It resides in the arrangement of matter. According to Weismann, 'The essence of heredity is the transmission of a nuclear substance of specific molecular structure.'[35] Only changes in this substance, or 'oscillations', are able to cause lasting changes in living beings. The whole mechanism of heredity, variation and evolution rests, not on the perpetuation of acquired characters through successive generations, but on the nature of a molecular structure.

Accordingly, at the end of the nineteenth century, two new elements appeared. Cytology revealed the existence of a structure with common properties in the cell nucleus; and critical analysis of the stability of species and of their variation revealed that heredity was attributable to the transmission of a particular substance. By common consent, this substance was placed in the chromosomes. Everything marked them out for this role: their constant numbers and shapes; the precision of their cleavage and distribution in cell-division; the reduction of their number to half in the germ cells; and finally, their fusion in the egg at fertilization, as a result of which the offspring received equal numbers of chromosomes from father and mother. Only the nuclear substance could carry the 'hereditary tendency'. And this tendency included not only the characters of the parents but also those of more distant forebears. Each of the germ cells united at fertilization contained chromosomes from the grandfather, great-grandfather, and so on. According to Weismann, the substance coming from preceding generations is present 'in proportion to their distance in time in an ever-diminishing ratio, according to the same calculation applied up to now by stockbreeders to the crossings of breeds in order to determine the fraction of "blue blood" contained in a descendant'.[36] The chromosomes of the father constitute one half of the child's nucleus; those of the grandfather one quarter, those of the tenth preceding generation $1/1024$, and so on. The problems of heredity fall in the domain of simple mathematics. At each generation, the chromosomes derived from the father and mother are reassorted. Statistical analysis makes it possible to evaluate the contributions from its various forebears to the hereditary substance of an individual. 'In biological phenomena,' said de Vries, 'differences from the mean follow the same laws as differences from the mean in all other kinds of phenomena controlled only by chance.'[37] Whether living or inanimate, all bodies obey statistical laws.

Only then could a science of heredity develop. It has often been said that Mendel's laws were 'rediscovered' at the turn of the century. What was rediscovered, however, was above all Mendel's approach, that of statistical mechanics: the same attention concentrated on a small number of characters with sufficiently striking differences for discontinuity to be introduced; the same method of enumerating the descendants of a cross, counting types and distributing them in finite classes; the same interest in populations rather than individuals, although individual pedigrees were recorded; the same statistical analysis of results; the same use of factorial symbols; the same distinction between the visible and the hidden. Hence, of course, the same phenomena, the same conclusions, the same laws. This approach to the analysis of heredity became so general that Mendel's work, which had been neglected for more than thirty years, was 'rediscovered' simultaneously in Germany, Austria and Holland and then in England, the U.S.A. and France. The rapid expansion of genetics from the very beginning of the twentieth century reflected its economic importance as much as its biological significance. Some of those interested in heredity were studying evolution; but they also included men who were attempting to increase production in agriculture and stockbreeding. The analysis of the mechanism of variation and the attempt to improve varieties of plants and animals involved the same methods, the same problems. Breeders and agronomists with extensive facilities at their disposal joined forces with biologists. They came together in certain societies, such as the Association of American Stockbreeders, whose president defined their aims as follows: the association 'has suggested that the scientists in biological lines turn for a time from the interesting problems of historical evolution to the needs of artificial evolution. It asks practical breeders, while seeking financial returns from breeding living things, to pause occasionally and study the laws of breeding. It has invited the breeders and the students of heredity to associate themselves together for their mutual benefit.'[38] From the start, therefore, the analysis of genetics was often applied to those organisms of importance in human affairs, such as wheat, maize, cotton or farm animals.

The qualities of the material played a major role in genetics. According to de Vries,

To study the general laws of heredity, complex cases must be completely excluded and the hereditary purity of parents taken as one of the first conditions of success. Furthermore, progeny must be numerous, since neither the constancy nor the exact proportions in cases of instability can be determined with a small batch of plants. Finally, to reach a definite choice of research material, it must be remembered that the main objective is to establish the relations that connect descendants to their parents.[39]

Botanists, such as de Vries, Correns and Tschermak, were the first to rediscover Mendel's attitude. Plants are, in fact, particularly suitable for the study of heredity: agriculture produces enormous populations; the processes of fertilization can be controlled. Studies were next extended to small laboratory animals, such as guinea-pigs, rabbits and rats. But this type of work called for an experimental object with exceptional qualities. It required an organism simple enough to be reared easily in the laboratory; small enough for large populations to be handled in limited space; and reproducing fast enough for successive generations to be studied over a short time span. Characters had to be easy to observe, mating frequent and fertility high. The cells of the organism had to be suitable for microscopic examination, and the number of chromosomes sufficiently small for their peculiarities to be noted. That *rara avis* existed: the fruit fly. For more than half a century, geneticists were to examine earnestly the eyes, wings and hairs of *Drosophila*, introduced by Morgan.

The techniques and methods adopted by genetics made it possible to study both the mechanism of variation in living beings and the characteristics of the structure underlying heredity. In the first place, by studying not just a few isolated individuals, but hundreds or even thousands of plants and animals, changes in characters of the population could be observed. During the second half of the nineteenth century, variations resulted from the progressive accumulation of minimal changes that, individually, often escaped the eye. Whether heredity was transmitted by extracts from each cell, as Darwin thought, or by a substance in the nucleus, as Weismann thought, it was the fluctuations to which each character was subject that provided the basis of variation and evolution. The intensity of a character was never exactly identical from one individual to another within the species. It could increase or decrease. In a population, some fluctuation always occurred; but the 'oscillations' of a character never diverged widely from the mean. Eventually, however, these differences accumulated and caused important variations when a sorting operation – performed deliberately by a breeder, or spontaneously by environmental conditions – always oriented selective pressure in the same direction. At the beginning of the twentieth century, the mechanism of character variation completely changed. For de Vries, it no longer occurred by a series of imperceptible modifications, but through sudden well-marked changes.

Species are not gradually transformed, but remain unaltered through successive generations. They suddenly produce new forms that are distinctly different from their parents and that are subsequently as perfect, constant, well-defined and pure as may be expected in any given species.[40]

Thus nature makes jumps. Her means of producing varieties and new species is mutation.

Unlike fluctuations and gradual imperceptible changes, mutations are accessible to observation and experiment. Provided that the material is suitable, the strain pure and the population large enough, one can measure their frequency, determine their character and establish the laws that govern them. These laws can be summed up by a few words: rarity, suddenness, discontinuity, repetition, stability, chance, generality. First of all, mutation is rare: during successive generations, the large majority of individuals are unaffected. 'The chance of finding a large number of mutants is small,' said de Vries. 'One must expect them to form a very low proportion of the culture.'[41] Mutant forms appear 'suddenly', 'without being expected'. From the outset, they have 'all the characters of the new type without intermediate .forms'; there is 'complete absence of transition between normal individuals and mutant forms'. The new forms have stable descendants; they persist through generations and are inherited in their turn; they show no 'tendency to return gradually to their original form'. The mutant forms are not produced once, but regularly, and 'the types are repeated in successive generations'. A mutation affects only one character at a time. Nevertheless, whatever the material and character studied, 'mutations are the rule'. Finally, there is no privileged direction in mutations, no connection between their production and the effect of external conditions, no correlation between their appearance and their usefulness. They occur at random and represent 'regression' just as often as 'progression'. They develop 'in all directions', said de Vries. 'Certain changes are useful, others harmful, but many are unimportant and are neither favourable nor unfavourable.'[42] Hence, all characters vary in all directions. They accordingly provide 'a very considerable amount of material to be sorted by the sieve of natural selection'.

The new status of variation justified *a posteriori* the empirical approach first adopted by Mendel and later by others for the analysis of heredity. The discontinuity, arbitrarily introduced for experimental purposes, in fact reflects the course of nature. If a single trait can assume several very distinct forms, if these forms can be represented by a series of symbols, this is because the determinant factor of the character can exist in several discrete states. Modification of the factor does not occur through a series of intermediates, but rather as an abrupt change from one state to another. Like changes in matter and energy, variations in heredity take place by quantum jumps. These jumps can be favoured and the frequency of mutation increased, by exposing the sperm of *Drosophila* to X-rays, as Muller did; or by treating the organisms with certain chemical compounds. Nonetheless, whether they occur 'spontaneously' or are 'artificially produced', mutations always appear at random. There is no relation whatsoever between their production and external conditions, no control exerted by environment. By definitively excluding the transmission of acquired characters, the analysis of mutations defined the respective roles played by heredity and environment in the formation of living beings. Environment can influence the organism only within the narrow limits of fluctuations permitted by what Weismann called the 'molecular structure of the hereditary substance', later known as the 'genetic material'. Outside these limits, there is no organism.

The other concern of genetics was to explore the organization and movement of what Mendel called 'factors', which were renamed 'genes' by the Danish geneticist, Johannsen. Experimentation gained access to these factors through hybridization. Crosses, however, were performed, not between different varieties of unknown origin, but between mutants of different characters derived from the same line. The actual mechanism of variation was then analysed. Like Mendel, the geneticists of the early twentieth century studied only a few characters at a time; these characters segregated independently. But as more and more different types of mutations were studied, anomalies began to appear. Certain groups of characters seemed to be 'coupled': they had a tendency to remain together through successive generations. Others, on the contrary, appeared to 'repulse' each other. In the *Drosophila* cultures of Morgan, there appeared a whole series of mutations which modified eye colour, or wing shape. In successive generations, these characters remained linked to the sex of the insects, as if by some invisible bond. It seemed to Morgan and his colleagues, Bridges, Sturtevant and Muller, that the genes were tied to some linear structure or 'linkage groups'. In *Drosophila*, genetics distinguished four linkage groups and cytology four chromosomes. It was possible to effect a convergence of genetics and cytology, by assigning each linkage group to a particular chromosome. It was even possible to associate the sex of the animal with one of these structures. Ultimately, the movement and distribution of chromosomes, and the exchange of genes between homologous chromosomes accounted for the hereditary differences between individuals of a species. By determining the frequency with which characters were united or separated in successive generations, it became possible to arrange them in linear order along the chromosomes like a string of beads. The relative distances between the genes could be estimated and a genetic map of the species drawn up.

For the geneticist, there are accordingly three ways of analysing heredity. Through characters, he can examine function; through their changes, he can examine mutation; through their reassortments he can examine recombination. Each of these methods enables him to reduce the genetic material into discrete units. Nevertheless, irrespective of the method of analysis, the end-result is the same: at one and the same time, the gene represents the unit of function, of mutation and of recombination. The material of heredity is thus resolved into elementary units that cannot be subdivided. Genes become the 'atoms' of heredity. Although through mutation a gene can appear in several discrete states, only one of them occurs in each chromosome. With its rigour and formalism, this quantum theory of heredity was not readily accepted by biologists, accustomed as they were to the everyday phenomena of continuous variations. It fitted in, however, with the concepts of physics, since the properties of the organism were thus reduced to indivisible units and their combinations subject to the laws of probability governed by chance. No more than the motion of an atom or an isolated electron, can one predict the particular combination of genes that will occur in an individual. When the dancer Isadora Duncan suggested to Bernard Shaw that they produce a child who should combine the mother's beauty with the father's intelligence, Shaw declined on the strength that the child might inherit his looks and her brains! Only on large populations is it possible to measure distributions and calculate probabilities.

Among all the constituents of living organisms, the genetic material has a privileged position. It occupies the summit of the pyramid and decides the properties of the organism. The other constituents are charged with the execution of the decision. Nevertheless, without the surrounding cytoplasm, the nucleus can do nothing. It is the whole cell that constitutes the elementary unit of the living organism, controls its properties, assimilates, grows and reproduces. The gene represents the ultimate in genetic analysis, but has no autonomy. Its expression usually depends on other genes. It is the genetic material as a whole, the particular combination of genes occurring in an organism, that determines its development, form and properties. Natural selection acts on populations by favouring the reproduction of certain individuals. Yet, as a result, it finally affects the genetic material itself in a roundabout way, by acting at three levels. First on the character, that is, the gene itself: any state is favoured, if it somehow makes reproduction more efficient. Next, it acts on the individual considered as an assortment of genes: certain combinations have more chance than others of producing descendants. Finally, it acts on the species, considered as the sum of all the genes belonging to all the individuals in the species: the appearance of new genes by mutation, or of new assortments by recombination, gives rise to new forms, from which natural selection takes its pick. Through a kind of cycle, the substrate of heredity also becomes that of evolution.

Classical genetics belongs to the field of biology which studies the organism as a whole or populations of organisms. It does not try to dissect the animal or the plant in order to recognize its components and study their function. The type of analysis used by genetics has been called the 'black box' method. The organism is considered as a closed box containing a large number of cog-wheels geared together in a very complex mechanism. Chains of reactions occur, intersecting and overlapping in all directions. One end of each chain lies at the surface of the box: it is the character. Genetics does not try to open the box and take the cog-wheels apart. It merely examines the surface to deduce the contents. Through the visible character, it attempts to find the invisible ends of the chains of reactions, to detect the structure that lies hidden in the box, controls its shape and properties. Genetics completely ignores the intermediate cog-wheels between the gene and the character. In the long run, a picture of great simplicity emerged from this type of analysis. It was simple in the mechanics imposed on the genetic material, as symbolized by the movement of the chromosomes, with their division, separation and reassortment. It was simple, too, in the structure itself, since the arrangement of the genes was represented by the most easily understood figure, the straight line. The gene itself, the hereditary element, appeared to be a three-dimensional structure of forbidding complexity, offering no hold for experiment. But to describe what underlies the forms, properties and function of the whole living organism, it was difficult to imagine a simpler model than a part of a string of beads. All the variations of character, all the mutations corresponded to changes in the nature or arrangement of the beads.

Within a few years, the gene theory transformed the picture of the living world. Taking everything into account, the properties and variation of animals and plants were based on the permanence and performance of a structure in the cell. Yet, even the black box method has its limitations. At the beginning of the twentieth century, it had allowed heredity to take shape, to be represented by a system of simple signs and to be treated mathematically. By ignoring the cog-wheels, however, it left a gap between the gene and the character. By means of symbols and formulae, genetics drew a more and more abstract picture of the organism. A product of reason, the gene seemed to be an entity with no body, no density, no substance. It then became a matter of conferring a concrete content on this abstract concept. The mechanism of heredity required the chromosomes to contain a substance endowed with two rare virtues: the power of reproducing itself exactly and the ability to influence the organism's properties by its activity. The aim of geneticists towards the middle of the century became to find the nature of this substance, to explain how genes act and to fill in the gap between gene and character. Still, neither the attitude of genetics nor its material and concepts lent themselves to this type of investigation. To gain access to the detailed structure controlling heredity, it was no longer enough to observe a few characters, and to measure the frequency of their association. A cooperation between genetics and chemistry was required.

Enzymes

In contrast to genetics, biochemistry belongs to that branch of biology which attempts to break down the organism into its component parts. During the second half of the nineteenth century, organic chemistry had marked the boundaries of its field. The next step was to define its position with regard to inorganic chemistry and to specify both the nature of the compounds and the mechanism of reactions peculiar to living beings. Until then, organic chemists had been occupied largely in attempting to identify and analyse the profusion of compounds they had isolated. All these substances had one common feature, the presence of carbon. They could, however, be classified according to a wide range of criteria: according to size – either large or small molecules; according to their nature – carbohydrates, fats or albuminous substances; according to the role they played – plastic or metabolic; according to their chemical function – alcohols, aldehydes, ethers, etc. The already long list was continually extended by the introduction of new substances, such as the phosphorus-rich acid isolated by Miescher at the very time when Mendel was hybridizing peas: its location in the cell nucleus earned it the name of 'nucleic acid', although no use could yet be found for it. Chemical analysis was generally limited, however, to isolating compounds from natural products, separating them from each other and altering them in order to unravel as subtly as possible the arrangement of the elements in the molecule. Although chemists could decompose organic substances, they did not yet know how to reconstitute them. Indeed they themselves long refused the very possibility of such syntheses. The transformations accompanying the flow of matter through a living organism set at naught the laws of inorganic chemistry. To move atoms and radicals so surely, to guide each element to its place in a molecule so precisely, to produce specific compounds so exactly, more than the laws of chemistry was needed: a vital force was required. Standing at the boundary between the living and the inanimate, organic chemistry put up a barrier, which it considered to be insuperable.

After the middle of the nineteenth century, the outlook of chemistry changed and the need to invoke some force beyond the laws of physics gradually decayed. One by one, most of the obstacles raised between organic and inorganic chemistry crumbled. First, the concept of energy and its conservation began to assume one of the roles previously ascribed to vital force. Energy exists in the very structure of a chemical compound: that is to say, in the forces linking atoms together in the molecule. When these bonds are broken and the atoms rearranged in a new structure with weaker bonds, excess energy appears in the form of heat, light, electricity or mechanical force. The energy contained in a compound can be calculated and the quantity of heat released by a reaction measured. When coal is burnt, for instance, the bonds linking carbon to carbon and oxygen to oxygen are broken; the two different types of atoms can then unite. But the energy contained in the carbon dioxide thus formed is weaker than the energy of the bonds joining the carbon atoms in coal. When an organism consumes glucose, only a fraction of the glucose is transformed into specific organic compounds. The rest is burnt and combines with oxygen, liberating not only carbon dioxide and water, but also energy. Energy can either be transformed into heat, or used again for other chemical reactions. In living beings, chemical transformations take place through the coupling of reactions which allow transfer of energy. In addition to the flow of matter through the organism, there runs a flow of energy. It is no longer vital force, but energy that is required for the formation of protoplasm and growth. According to Helmholtz,

There may be other agents acting in the living body than those agents which act in the inorganic world, but these forces, insofar as they cause chemical and mechanical influence in the body, must be quite of the same character as inorganic forces . . .; there cannot exist any arbitrary choice in the direction of their actions.[43]

From thermodynamics arose a physical chemistry that calculated the available energy in compounds, determined the rates of reactions and measured their equilibria. The rules of inorganic chemistry were gradually extended to organic compounds. In a whole series of biological phenomena, the rules of chemical equilibrium and 'mass action' were found to hold true. In living organisms and in the laboratory, the laws of chemical dynamics are the same.

The second thread that helped to reunite organic and inorganic chemistry was the total synthesis of organic compounds. In the laboratory, there are several ways of producing a compound. It can be obtained by modifying other substances: either splitting a more

complex compound into fragments or adding elements or radicals
to a simpler compound. Alternatively, the whole architecture of the
molecule can be built up from the constituent elements alone. For
the chemist, only the latter method constitutes total synthesis. The
very possibility of applying this method to the constituents of living
beings had previously seemed to be ruled out. In organic compounds, elements occur in such a limited number, but in such exact
proportions and such varied combinations that these substances had
appeared to lie beyond the reach of laboratory synthesis. All previous
efforts to reproduce the work of nature by the art of the chemist had
failed. Certainly, Wöhler had succeeded in synthesizing urea and
oxalic acid and Kolbe had synthesized salicylic and acetic acids. Such
synthesis, however, involved special reactions, not a general method
for producing a series of compounds. Furthermore, in all these cases
it was necessary to start from a compound that was already a carbon
derivative. Unable to link together carbon and hydrogen, chemists
considered the barrier between organic and mineral to be insuperable. Only vital force could overcome the counterflow of the forces
acting on matter. According to Liebig, the organic chemist was not
even under the obligation to verify the results of organic structural
analysis by synthesis.

After the middle of the nineteenth century, the question of organic
synthesis presented itself in different terms. For Berthelot, it became
'necessary to form organic compounds from their elements, particularly those compounds possessing particular functions unlike those
known to inorganic chemistry'.[44] It was therefore no longer a
question of obtaining a few compounds by exceptional means. The
aim was to perfect a method of synthesizing the most varied kinds
of organic compounds and producing the entire range. This was
possible because organic chemistry was based on the characteristics
of carbon, and the properties of its derivatives on their chemical
functions. 'Organic compounds,' wrote Berthelot, 'can be classified
according to eight functions or fundamental types, which comprise
all the compounds known to-day and all those we can hope to
obtain.'[45] These eight chemical functions can themselves be arranged
in several groups according to the number of elements associated

with the carbon. There is thus a first class formed of compounds with only two elements, the hydrocarbons; another class of compounds with three elements, carbon, hydrogen and oxygen, grouping four chemical functions, alcohols, aldehydes, acids and ethers; then a class of nitrogenous compounds in which two functions are represented, alkalis and amides; lastly, there is a function called 'compound metallic radicals', containing metals bound to certain ethers. The order of complexity imposes the order of the syntheses. For the main difficulty lies in the very first stage; that is, making carbon form new links with other elements, particularly with hydrogen. Once hydrocarbons have been formed, all the other functions can be derived by synthesis. Linking carbon with other elements, however, is no longer a purely empirical operation. It has a theoretical foundation based on the concept of valency. For Kekulé, what distinguishes carbon and makes it unique in the living world is its 'tetravalency'. Each atom of carbon can form four bonds with other atoms, bonds that are 'saturated' or 'unsaturated' by other elements. Six atoms of carbon can be joined two by two to form a chain that closes in a 'ring' or 'aromatic nucleus'. The tetravalency of carbon, therefore, makes it possible to define the relative position of atoms in a compound, to characterize the bonds that form between them and to explain isomers by the distribution of atoms in space – in short, to represent any given organic molecule by a system of symbols and predict its chemical properties. From there, it is possible to deduce the general laws controlling hydrogenation of carbon and the proportions of elements to be used to obtain a given compound. Under the influence of an electric arc or of heat, carbon combines directly with hydrogen to produce the simplest hydrocarbons such as acetylene or ethylene. By a series of substitutions, all the hydrocarbons can then be synthesized step by step. 'These methods are general,' wrote Berthelot, 'and allow all hydrocarbons to be formed from their elements: they therefore establish the final link between organic and inorganic chemistry, both proceeding by the same principles of molecular mechanics.'[46] Hydrocarbons represent the carbon skeletons onto which all the other chemical functions can be grafted; either directly, by transforming a hydrocarbon into an alcohol, aldehyde, acid, and so on; or indirectly, by first forming an alcohol which is then changed into an aldehyde or acid, etc. Thus, using only the elements, the effect of 'chemical affinities', and such physical forces as electricity or heat – that is, by laboratory methods alone – it is possible to produce a multitude of natural organic compounds. 'By the fact of this formation and by imitation of the mechanisms operating in plants and animals,' said Berthelot,

it can be established that, contrary to former opinions, the chemical attributes of life are due to the action of ordinary chemical forces in the same way as the physical and mechanical attributes of life result from the action of purely physical and mechanical forces. In both cases, the molecular forces operating are the same, since they produce the same effects.[47]

Chemists, however, were not satisfied with imitating nature and reproducing its compounds. They were also able to create novel compounds that resembled natural products and shared some of their properties. This gave some concrete support, in a way, to the abstract laws of chemistry. One had no longer only to imagine the transformations that might once have occurred in the chemistry of living beings. 'We can claim . . .,' said Berthelot,

to conceive of the general types of all possible substances and to bring them into being . . . to form anew all the materials which have been developed since the origin of things, under the same conditions and by the same laws and same forces that nature used in their formation.[48]

There was no longer any theoretical limit for organic chemistry.

All distinctions between reactions in the living world and in the laboratory were finally to disappear, following a surprisingly wide detour that led chemistry to intervene in a domain previously reserved to naturalists. It was, in fact, chemistry that revealed the role of micro-organisms in this world. It was through the use of chemical methods that the last traces of spontaneous generation were eliminated. Until then, organic substances had been characterized by their composition and chemical properties. At the end of the nineteenth century, the molecular structure, the relative position of atoms, became more important, since certain compounds, known as isomers, although of identical chemical composition, were found to possess different properties. The chemical species, said Pasteur, 'is the collection of all the individuals which are identical in nature, proportion and arrangement of their elements. All the properties of compounds are a function of these three factors.'[49] Certain optical characteristics of compounds could be related to their 'molecular asymmetry', a point which provided analysis with another tool. It soon proved impossible, however, to produce, by laboratory reactions, the asymmetry found in natural products. 'All the artificial products of laboratories,' said Pasteur, 'have superposable images. On the contrary, most natural organic products . . . those that play an essential part in the phenomena of plant and animal life, are asymmetric.'[50] The compounds found in organisms, therefore, have a property distinguishing them from the same ones made in the laboratory. In every organism, there is an unknown force, not reproducible in the laboratory, that produces asymmetry in chemical activities. By the study of this asymmetry, chemistry penetrated the world of microscopic organisms, through the intermediary of fermentations. All fermentation involves two factors, one passive, the other active. The first factor, such as sugar, is said to be 'fermentable'; it is transformed under the influence of the second factor, or ferment, a nitrogenous substance of 'albuminoid' nature. For Liebig, fermentation was the property of certain organic substances that, being themselves in a state of 'metamorphosis', were able to transmit this property to neighbouring substances. For Berzelius, it was a 'catalytic' property that gave a substance the capacity to transform another one without the catalyst itself being a part of the transformation. In every case, albuminous substances were assigned by some mysterious force the role of ferment – that is to say, the capacity to act by contact on fermentable compounds. In all cases, the power to ferment was not a property of the organism as a whole, but of certain of its constituents. For Pasteur, it was a completely different matter. If living beings introduced molecular asymmetry into chemical reactions, then conversely molecular asymmetry was a sign of the presence of a living organism. The normal procedure in science was thus reversed. It usually proceeds from theoretical knowledge to practical questions of interest to man. Here, it took the opposite direction. It was through the difficulties encountered in the beer, wine and alcohol industry that Pasteur found a way of associating closely biology and chemistry. Deviations in fermentations, 'diseases' of beer and wine, caused the formation of asymmetric compounds. They were, therefore, connected with the presence of living organisms. For Pasteur, the abnormal, the pathological does not provide a model for physiology. It gives a basis for experimentation. It points out the phenomenon that investigation transforms into a physiological process. Anomalies in fermentation simply become other types of fermentation. Whether fermentation is alcoholic, amylic, 'viscous', acetic, lactic, butyric and so on, it is always associated with the multiplication of microscopic organisms. 'Real ferments are organized beings,' said Pasteur.[51] Moreover, in each kind of fermentation, there is found a particular type of organism that can be isolated, cultivated and studied. For a given substrate, the specificity of the organism involved determines the specificity of chemical reactions and therefore of fermentation. Not that one substance cannot be fermented by several different organisms, or that a given organism cannot ferment many different substances. In fact, a whole range of compounds is produced by fermentation; it is the whole spectrum that characterizes the organism. According to Pasteur, 'To every fermentation may be assigned an equation in a general sort of way, an equation, however, which in numerous points of details is liable to the thousand variations connected with the phenomena of life.'[52] Microbial fermentation is like animal nutrition: both reflect chemical activities of the living organism.

How greatly this approach differs from what preceded it! Not only did it modify the nature of the bonds between biology and chemistry; it also changed the general picture of the living world, the relations established between living beings, the distribution of roles in the chemical activities on earth. Suddenly, the invisible world, revealed in the late seventeenth century by the microscope, but hitherto unused and almost ignored, found a place, a status, a function. Pasteur's outlook had two facets, which he developed in parallel. On the one hand, the specificity of the micro-organism determined the nature of fermentations, as cause produces effect. The concept of specificity then extended to an unexpected field, pathology: a whole series of human and animal diseases became the result of the invasion of the organism by a particular 'germ'. Conviction of the correctness of this principle was so strong that it applied even in cases where the responsible agents – later to be known as viruses – could neither be seen nor cultivated in a test-tube. On the other hand, correlation was established between the chemical effects on external substances or organisms and the living character of the agent involved. The problem of fermentation could then be reversed and the question expressed in a new way. 'One of two things may happen,' said Pasteur:

as the ferments in fermentations proper are organized, if oxygen alone, as oxygen, produces them by contact with nitrogenous materials, then they are spontaneous generations; if these ferments are not spontaneous beings, then this gas takes part in their formation, not as oxygen alone but as a stimulant of a germ introduced at the same time as the gas, or already existing in the nitrogenous or fermentable materials.[53]

By repeating and perfecting Spallanzani's old experiments, by performing them with the chemist's precision, Pasteur finally excluded any possibility of spontaneous generation: even with microscopic organisms, the living is born only of the living. Wherever bacilli are found, an identical bacillus has previously existed to engender them.

But if the demon of spontaneous generation had at last been exorcized, the demon of vitalism remained as firmly implanted as ever. In fact, the late nineteenth century was faced with a paradox: on the one hand, organic and physical chemistry had denied a special status to the chemistry of organisms; on the other, crystallography had revealed a special property peculiar to the constituents of living organisms, and microbiology considered fermentation to be a property of the living cell. This led to endless polemics, infused with all the passion that the nineteenth century was wont to display. For the existence of soluble 'diastases' had long been recognized: substances that mediated the decomposition of certain sugars and albuminous compounds in the test-tube, in the absence of a living organism. It was, therefore, necessary to distinguish two classes of ferments: 'organized' and 'unorganized'. But this problem was resolved by the chemists at the very end of the nineteenth century. They were able to crush cells, to prepare cellular extracts and to look for ferments. By grinding a cake of dried yeast with sand, Büchner found that 'yeast juice', cleared of all living cells by filtration, was still able to convert glucose to alcohol.

> For the production of the fermentation process, no such complicated apparatus is necessary as is represented by the yeast cell. It is much more likely that the agent of the juice which is active in fermentations is a soluble substance, doubtless an albuminoid.[54]

All known ferments thus behaved in the same way. All were substances, not living beings. All acted outside the organism. The work of Büchner opened the way to the discovery of many other ferments, able to catalyse specifically a wide variety of reactions. Furthermore, with the development of physical chemistry, catalysis lost much of its mystery. The chemists learnt how to measure the parameters of a chemical reaction, its rate, equilibrium and reversibility. Catalysis affects only one of these parameters: it increases the rate, just as does a rise in temperature. In Ostwald's words,

> Catalysis is the name applied to processes in which the rate of reaction is changed by the presence of a substance which, at the end of the reaction, is in the same state as at the beginning. These substances only change the rate of the reaction; they do not take part in it.[55]

Catalytic phenomena are by no means restricted to the chemistry of life. There are inorganic substances, heavy metals, for instance, such as platinum black, which because of their enormous surface catalyse a whole series of reactions. The diastases differ from such inorganic catalysts principally by their specificity: they mediate only one single reaction. Thus the obstacle which, since Lavoisier, had limited investigation of chemical reactions in living organisms disappeared. The processes taking place in the organism became subject to the laws of chemical dynamics.

Büchner's demonstration of the conversion of glucose to ethanol by cell-free extracts was important not simply for the new light it shed on the chemistry of the living organism, but above all because it provided a new method of analysis. With tissues and whole cells, it is often difficult and sometimes impossible to make certain compounds penetrate the cell membranes. With cell extracts, on the contrary, it becomes relatively easy to analyse a reaction: compounds suspected of taking part in it can be added, or removed; the effects of possible inhibitors can be examined. It is no exaggeration to say that since that time the investigation of cell-free extracts has remained the chief method by which chemists study living organisms. Thus, at the beginning of the twentieth century, a new branch of chemistry was established: biological chemistry or biochemistry. Organic chemistry continued to explore the ensemble of carbon-derivatives, to study their properties and produce new compounds by synthesis. Biological chemistry, on the other hand, investigated the constituents of living organisms and their transformations in relation to biological function. Situated at the very heart of biology, biological chemistry is related to all the other biological disciplines. It differs from them, however, by its methods, its objects and even its approach to the organism. When the organization of a living body is destroyed, life disappears, but by no means all its manifestations. Disintegration of the organism arrests certain phenomena such as reproduction and growth; but it allows others, such as fermentation, to continue. According to Loeb, the role of biological chemistry consists, therefore, in 'distinguishing the functions which depend only on chemical constitution from those that also require a particular physical structure of the living substance'.[56]

Two currents can be distinguished in the early development of biological chemistry. The first sought to define the chemical nature of the cell and to analyse the 'protoplasm' in physico-chemical terms. At that period, the limit of structural analysis was set by the resolving power of the optical microscope. Such structures as the nucleus, the cell membrane, the mitochondria, etc. could be seen in the cell. 'Protoplasm' did not appear to have real structure; it was a kind of emulsion, a suspension of granules or 'miscellae' in a liquid, what was called a colloid. According to Loeb, 'The substances constituting living matter, whether they are liquid or solid, are colloids.'[57]

Colloids, as opposed to crystalloids, are not an attribute of living organisms alone: they can be prepared in the laboratory by making a fine suspension of gold or platinum particles in water, for instance Such a suspension exhibits the particular qualities of stability surface area and electric charges that favour chemical reactions and contribute to catalysis. Albuminous compounds and fats extracted from various organisms easily produce colloidal solutions. In the end, besides the variety of structures visible to the naked eye or through the microscope in living organisms, it is the colloidal nature of protoplasm that gives cells their special character. 'Life depends on the maintenance of certain colloidal solutions,' said Loeb, 'all agents that cause general gelation also bring life to a standstill.'[58] That is what happens when albumins are coagulated by heat or the action of heavy metals. By destroying visible structures but respecting the colloidal nature of protoplasm, it ought to be possible to investigate protoplasm. At the beginning of the twentieth century, however, the means which biochemistry had at its disposal were still insufficient. Only with the development of physical methods, particularly ultracentrifugation, did it become possible to interpret the contents of cells not in terms of colloids, but of molecules.

The second current in biochemistry sought to study cell constituents and reactions, following the path indicated by Büchner. The first point of attack was the determination of the different stages in the breakdown of glucose by yeast. Very soon, however, such studies were extended to other reactions; this type of biochemical analysis made great strides at the beginning of the present century. Its chief method consisted in carefully dissociating organisms, tissues or cells and 'opening' them as gently as possible in order to gain access to the cell contents. Once a process mediated by such an extract had been identified, biochemists tried to define the constituent reactions, to isolate the compounds involved and to purify them by laboratory methods. The goal was to reconstitute the entity that had been destroyed, to reunite what had been separated, to develop a 'system' of which it was possible to study the properties, measure the parameters and define the requirements. The reaction could then be represented by the symbolism of chemistry. This approach, essentially reductionist, sharply distinguished biochemistry from other biological disciplines. Indeed, other biologists often criticized the biochemist for studying objects which no longer had anything in common with living organisms, for creating artefacts, for trying to explain the whole in terms of its parts: in short, for drawing false conclusions from his analyses. Although to some extent armed against such criticism, the biochemist did nevertheless try to parry the blows; at each step of his analysis, he compared the phenomena observed in the test-tube with those which take place in the organism.

The biochemists used either animal tissues or cultures of micro-organisms to prepare their extracts. There again, certain objects, such as rat-liver, pigeon-muscle or suspensions of yeast provided the most attractive material because of the ease with which they could be obtained and handled. In such extracts, biochemists tried to identify molecules and define reactions. The substances composing living tissues could be assigned to three major chemical categories: sugars or carbohydrates, fats or lipids, albuminoids or proteins. In each class, there existed large and small molecules. Unstable and difficult to prepare, to isolate and to characterize, the large molecules, particularly proteins, were not yet amenable to investigation. Neither the techniques nor the concepts of that time were appropriate. The small molecules, in contrast, lent themselves to the existing methods of organic chemistry. It was possible to purify and analyse them, to study their properties and follow their transformation during metabolism. In many cases, they could even be synthesized. Consequently, the biochemists recognized an increasing number and variety of such molecules and of reactions which they underwent in the organism. In the laboratory, these reactions take place very slowly at body temperature. But for each reaction, a particular biological catalyst was found, a diastase, or 'enzyme' as it was hence-forth to be called, that increased the rate several thousand fold. The kinetics and properties of enzymes were investigated and defined. Little by little, for each known reaction the corresponding enzyme was detected. Each enzyme was designated by the name of its sub-strate, to which the suffix '-ase' was added. There were thus enzymes for the degradation of each class of compound – sucrases, lipases, proteases. Within each class, sucrases, for instance, there were special enzymes for each type of carbohydrate, an amylase, a lactase, a saccharase, and so on. There were also enzymes, like maltase, that could not only degrade a sugar, but under certain conditions resynthesize it from the products of its degradation. There were even enzymes responsible for the processes of respiration. Since the work of Lavoisier, respiration had been considered a special type of combustion that took place slowly, at body temperature. At the beginning of the twentieth century, respiration became the result of specialized enzymatic activities that catalyse the slow oxidation of foodstuffs in small successive stages. In fact, foodstuffs are first digested; then the products of digestion are oxidized by the withdrawal of hydrogen atoms. Oxidations and reductions are coupled by means of small molecules that can be alternatively oxidized or reduced at great velocity. Respiration is then transformed into a series of oxido-reduction reactions, each catalysed by an enzyme: electrons are transferred along a chain that begins with metabolites and ends with molecular oxygen. In fermentation, which takes place in the absence of oxygen, the latter is replaced by certain organic compounds. Always, however, is a particular enzyme found that catalyses a particular reaction. The specificity of catalytic action in the chemistry of living organisms was so high that it could be described by the aphorism: one reaction, one enzyme. Conversely, once their action had been defined, enzymes provided chemistry with a new weapon for analysis and synthesis. By means of the enzymes he had characterized, the biochemist could handle the small molecules of the cell as he wished, reducing or enlarging them with precision, removing an atom here, adding a radical there. The biochemist thus acquired a sureness and technical skill that could hardly have been predicted.

Thus armed with new material and methods, biochemistry developed a whole series of new concepts. First, the discovery of compounds and reactions in an ever-increasing number led to the notion of 'intermediary metabolism', defined as the sum of all the reactions by which nutrients are transformed into specific compounds. It had long been clear that a nutrient does not contain all the compounds that constitute an organism or a cell. It therefore had to be degraded and specific compounds constructed from the products thus formed. From the time of Pasteur, this was revealed by observations on the growth of micro-organisms: yeast, for instance, can multiply in chemically defined media containing mineral salts and a single organic compound, such as glucose, as the source of carbon and energy. Once it has entered the cell, the glucose has to be rearranged chemically to produce all the compounds indispensable for the growth and life of yeast. These transformations do not occur simultaneously. They can be broken down into a number of steps, each simple and accessible to analysis. They therefore correspond to chains of reactions, involving a whole series of intermediate compounds that usually play no special physiological role, each being the product of one reaction and the substrate of the next. The nutrients are first attacked by specific enzymes that degrade them, fragment them and transform them into small molecules. In their turn, the small molecules serve as substrates for other enzymes which remodel them, add atoms, substitute radicals, lengthen and link them – in short, which manufacture the characteristic constituents of the organism. The organism thus becomes a sort of chemical factory swarming with multitudes of small molecules formed from foodstuffs by chains of degradation and then transformed into specific compounds by chains of synthesis. Furthermore, the same metabolic chain is often found in different organisms. For example, the breakdown of glucose during its fermentation by yeast and during the contraction of a muscle in the absence of oxygen involve the same reactions and intermediates. Thus became discernible the unity of chemistry in the living world.

Nutrients have to provide the organism with energy as well as with building materials. When a yeast uses a sugar for growth, whether in the presence or absence of oxygen, whether by respiration or fermentation, only part of the sugar consumed is converted into constituents of yeast. The remainder supplies the energy necessary for its work. To increase and multiply, to maintain the order of the living world despite the tendency of the universe to decay, organisms must be provided with energy from an external source. In the long run, it is the sun that provides energy for most living beings. Individual organisms, however, have different means of assuring their energy supply. Some, such as green plants, draw their energy directly from sunlight by photosynthesis; others, such as certain bacteria, obtain energy from the oxidation of inorganic compounds; while others again, like most animals, procure it from the oxidation of organic compounds. But in every case, to be available when required, energy must be stored in chemical form. Biochemists found that it is stocked in certain phosphorus compounds containing the so-called 'energy-rich bonds'. Through the formation, synthesis and transfer of such bonds the energy of biological systems is stored, released or exchanged. In the last analysis, a single unique compound common to the whole living world, adenosine triphosphate, constitutes the energy storage in all organisms. Whether in bacteria or mammals, whether energy is obtained by respiration or fermentation, the degradation of a sugar always occurs through a series of similar operations; the same stages and same reactions lead to the production of the same energy-rich compound. This strengthened the concept of the unity of function in the living world.

It also became possible to analyse the nutrition of organisms, to define their food requirements and to determine what is indispensable for growth and reproduction. Certain compounds, called 'vitamins' appeared to be necessary for the health and life of mammals. Others, named 'growth factors', were required for the multiplication of certain microbes. By trying to define the kind of compounds required by organisms as different as bacteria and mammals, physiologists and biochemists observed curious analogies: the 'growth factors' necessary for bacteria were often identical with the 'vitamins' indispensable to mammals. Moreover, these compounds were not found only in those organisms which need them as nutrients, but in all organisms. Some organisms are themselves able to synthesize all the compounds, others not. The latter, therefore, have to be provided ready-made with the components they cannot themselves produce. Hence the notion of 'essential metabolites' necessary to the life of all organisms. It was no longer only the functioning of organisms that expressed the unity of the living world. It was also their composition.

In the first part of the present century, the chemistry of living organisms thus became accessible to experimentation. Hundreds of reactions were studied in test-tubes. A considerable number of relatively simple compounds were analysed. The transformations that provided energy reserves and building materials were followed. The more closely these reactions became defined, the less easily could they be distinguished from those carried out in the laboratory. The originality of the chemistry of living beings lay chiefly in the enzymes. It was because of the specificity, precision and efficiency of enzymatic catalysis that the network of all chemical operations could operate within the minute working-space of the cell. It was the high degree of selectivity allowing each enzyme to choose only one optical isomer of a given compound, that impressed asymmetry on the chemistry of organisms. By attempting to characterize the enzymes and to determine their nature and mode of action, bio-chemists gradually came to associate enzymatic activities with the presence of proteins. Eventually, each particular enzymatic activity became the attribute of a particular protein. If the chemistry of living beings had a secret, then it must lie in the nature and properties of proteins. But although biochemical methods were perfectly adequate for the analysis of relatively simple molecules, they proved quite unsuitable for analysis of these enormous molecules. The labile protein molecules easily denatured. Difficult to handle, they could not be studied by the traditional techniques. Upon chemical degradation, they yielded a few simple molecules, the 'amino acids', hundreds of which appeared to be linked to one another in one protein molecule. Gradually new methods were elaborated for preparing, isolating and purifying proteins. It even became possible to crystallize certain enzyme-proteins. Another barrier between the two types of chemistry was thus swept away. But proteins remained exceptionally complex structures. Biochemistry had not yet found a route of access to the structure of such large molecules, to the arrangement of amino acids which enabled them to bind specifically one particular molecule and no other, and to catalyse its transformation. The analysis of proteins required new techniques and concepts. Only in the middle of the present century was a suitable methodology to emerge, a product of the combined influence of physics, polymer chemistry and information theory.

*

At the beginning of the twentieth century, biology took a new turn, under the impetus of two new sciences, genetics and biochemistry. Firstly, because they both introduced a quantitative rigour hitherto unknown in biology: it was no longer sufficient simply to notice the existence of a phenomenon; henceforth, it became necessary to estimate its parameters – to measure reaction velocities or recombination frequencies, to determine equilibrium constants or mutation rates. Secondly, both genetics and biochemistry changed the centre of gravity of living bodies. Organisms were no longer thought of simply as organs and functions arranged in depth; they no longer appeared as curled round a source of life from which organization radiated. For biochemistry, the activity of the organism was dispersed throughout the cell, in the thousands of colloidal droplets where chemical reactions took place and structures were built up. For genetics, this activity was concentrated in the cell nucleus, in the movement of chromosomes where forms were decided, functions ordered and the species perpetuated. Each science referred to its own model. On the one hand, chemists spoke of molecular structures and enzymatic catalysis; they explained how organisms drew their energy from the environment and thus counteracted the natural degradation of energy: there was no longer just a flow of matter through the organism, but also a flow of energy. On the other hand, geneticists described the anatomy and physiology of a third-order structure placed in the chromosomes; they ascribed the memory of species to the fixity of this structure and the appearance of new species to its changes. The properties of living organisms were ultimately based on two new entities: what biochemists called protein and what geneticists called gene. The first is the unit which carries out the chemical reactions and gives living organisms their structure. The second is the unit of heredity that controls both the reproduction and the variation of a function. The gene gives orders. The protein executes them.

Towards the middle of the century, genetics and biochemistry found themselves at more or less the same point. Both had succeeded in detecting the unity of action at the centre of its own domain. Both knew, therefore, the object to be investigated. But both lacked the means necessary to succeed. In fact, before the Second World War, biology had become a rather partitioned science. Each specialist grappled with his own problems, using his own materials. In the same institute, often on the same floor, could be found two colleagues, one working on genes and the other on molecules. The conclusions reached by genetics required the presence in the chromosomes of a substance capable of very unusual actions: on the one hand, it had to determine the structures and functions of living organisms; on the other, to produce exact copies of itself, without excluding the possibility of rare variations. Chemistry had found two kinds of substance in the cell nucleus: proteins and that acid which Miescher had named 'nucleic acid' in the previous century. But the structure of nucleic acid was still almost unknown. It was composed of four particular molecules, two 'purine bases' and two 'pyrimidine bases', each linked to a sugar and a phosphate group to form a 'nucleotide'. The four compounds were associated to form a 'tetranucleotide'. Nucleic acid thus seemed to be a molecular species without variety or fantasy, and therefore inapt to play the slightest role in heredity. This role was therefore attributed to proteins, although their properties seemed hardly suitable for the task. By its complexity, heredity appeared to lie beyond the range of experimental chemistry. According to J. S. Haldane:

The more we discover as to physiological activity and inheritance, the more difficult does it become to imagine any physical or chemical description or explanation which could in any way cover the facts of persistent coordination.[59]

By the end of the nineteenth century and the beginning of the twentieth, there was nothing left of the old form of vitalism, the vitalism which early biology had had to postulate in order to acquire independence. With the development of experimental science, of genetics and biochemistry, it was no longer possible, except for the mystic, seriously to invoke some principle of unknown origin, an x

eluding the laws of physics by its very essence, in order to account for the existence and properties of living organisms. If physics did not seem able to explain all the phenomena of life, this was no longer because of a force peculiar to the living world and beyond the reach of all knowledge; it was because of the limitations inherent in observation and investigation and because of the complexity of living organisms as compared with inanimate matter. Just as certain characteristics of atoms could not be reduced to mechanics, so certain special qualities of the cell could perhaps not be interpreted in terms of atomic physics. 'The recognition of the essential importance of fundamentally atomistic features in the functions of living organisms is by no means sufficient for a comprehensive explanation of biological phenomena,' said Niels Bohr.

The question at issue, therefore, is whether some fundamental traits are still missing in the analysis of natural phenomena, before we can reach an understanding of life on the basis of physical experience . . . On this view, the existence of life must be considered as an elementary fact that cannot be explained, but must be taken as a starting point in biology, in a similar way as the quantum of action, which appears as an irrational element from the point of view of classical mechanical physics, taken together with the existence of the elementary particles, forms the foundation of atomic physics.[60]

What could impose a limit on understanding the living world was, accordingly, no longer a difference in nature between the living and the inanimate worlds. It was the inadequacy of our means and even of our possibilities of analysis. Moreover, in living organisms, the complexity of the constituents had nothing in common with that of the molecules studied by classical physics and chemistry. Far from eluding the laws of physics, living organisms, as Schrödinger said, might perhaps involve 'other laws of physics hitherto unknown, which, however, once they have been revealed, will form just as integral a part of this science as the former'.[61] The question, therefore, was no longer whether a mysterious force was required to justify the origin, properties and behaviour of living beings. It was whether the laws already discovered in the analysis of matter were enough in themselves, or whether new laws had to be found. To become a science, biology had had to cut itself off radically from physics and chemistry. To continue investigation of the structure and functioning of living beings in the middle of the twentieth century, biology had to cooperate closely with them. From this union, molecular biology was to be born.

5

The Molecule

In the middle of this century, the status of organization changed once more. It became the structure of constituent parts that determined the structure and integration of the whole. Organization was buried deeper in living organisms and came to reside in the smallest details of the cell. Until then, and despite the presence of a nucleus and various organelles, the cell had appeared to be a sort of 'bag of molecules'. If countless chemical reactions could dovetail in the cell, if catalysis was possible, it was mainly because of the nature of protoplasm, that ill-defined colloidal mesh. For coordinating the activity of organs and tissues, complex organisms had special equipment at their disposal. Nerves and hormones wove a network of interactions through the body, by which even the remotest elements of the organism were linked. The unity of organization was based on the existence of specialized mechanisms that regulated functions. Nothing similar operated in simpler structures.

With the development of electronics and the appearance of cybernetics, organization as such became an object for study by physics and technology. The requirements of war and industry led to the construction of automatic machines in which complexity increased through successive integrations. In a television set, an anti-aircraft rocket or a computer, units are integrated which already result themselves from integration at a lower level. Each of these objects is a system of systems. In each of them, the interaction of the constituent parts underlies organization of the whole. There can be integration only in so far as the parts communicate with each other and reciprocally modulate their own particular activities in terms of the over-all objective. Until then, the coordination of components was considered as a property that existed only in certain systems. Thereafter, organization and integration of the components were inseparable. Each became the very condition for the existence of the other, both its cause and effect. There is integration only in so far as components react to each other. There is reciprocal influence between the constituents only in so far as the system is an integrated one. If exchanges are possible between the constituents of an organized body, it is because their structure so lends itself. But at the same time, the organization of these elements contains the potential sequence of their future arrangements and thus of its own transformations. In the last analysis, coordination of activities determines not only the properties of an integrated system, but the way it evolves. It is from the relation between structures and functions that the internal logic of such a system is born.

The qualities, functions and development of a living organism thus simply express the interactions between its components. Underlying each character are the properties of certain structures. The analysis of functions cannot be dissociated from that of structures: structure of cells for the functions of the body; structure of molecules for the functions of the cell. But to interpret biological processes in terms of the molecular structures which characterize the cell requires a convergence of analyses as well as a combination of methods. For a century, experimental biology had progressively split into a series of branches which tended to become more and more isolated. Each discipline rested on a small number of techniques which defined the limits of its special field. In the middle of the century, however, the different disciplines found themselves forced together. Progress had become dependent on a unified attack, an articulation of view-points, an adaptation of methods: in short, the creation of a 'molecular biology'. To do molecular biology, it was no longer enough to use one technique, to investigate all the parameters of one particular phenomenon. It became necessary to exploit all the means available in order to define the architecture of the compounds involved and the nature of their relations. Attempts were no longer made to study separately genes, chemical reactions and physiological effects. The chain of events leading from the gene to the character had to be described in terms of molecular species, of syntheses and of interactions. The organization of a macromolecule, the 'message' formed by the arrangement of chemical patterns along a polymer chain, held the memory of heredity. It became the fourth-order structure that determined the form, properties and functioning of a living organism.

Biochemistry and physics, genetics and physiology thus fused into a single discipline, that is to say, molecular biology. This science could no longer be the concern of isolated scientists, each preoccupied with a particular problem and a particular organism. It required a combined effort, human and technical. In a single institute, in a single laboratory, a cooperative attack was initiated by specialists who, although separated by their backgrounds, were united in a common enterprise, using a common material. There were no longer two kinds of biology, one interested in the whole organism and the other in its components. There were simply two aspects of the same object. Molecular biologists still examined the black box from the outside, in order to observe its properties; but at the same time they opened the box in order to detect the cogwheels, to dismantle them and to try to reconstruct the mechanism from the separate parts. Organism or constituents, each could be interpreted only with reference to the other. Previously compelled to isolate itself in order to define its aims and methods, biology was now brought into close association with physics and chemistry; an association which detracted nothing from its own special character.

Macromolecules

By the middle of the nineteenth century, the notion of energy and its conservation had changed the picture of the living world; firstly, by establishing a link between the chemistry of beings and that of things; secondly, by giving a common basis to all the most varied activities of the organism. Energy had thus replaced vital force in some but not all its functions. There are billions of cells in a complex organism, millions of molecules in a cell; but there was no way to explain the specificity of structures, the arrangement of the cells, or the positioning of atoms in isomers, those substances of identical composition but of different properties. Statistical mechanics made it possible to interpret the average behaviour of large populations of molecules. Genetic analysis, however, revealed that biological properties were not the result of statistical molecular events; but that, instead, they were based on the quality of some substances contained in the chromosomes. In contrast to the order of inanimate bodies, the order of living organisms could not be extracted from disorder. It depended on the reproduction of an already existing order. According to Schrödinger, 'Life seems to be orderly and lawful behaviour of matter, not based exclusively on its tendency to go over from order to disorder, but based partly on existing order that is kept up.'[1]

In the middle of the nineteenth century the concept of information opened the way to the investigation and transmission of this order. By ignoring individual events and considering only the average behaviour of a population, statistical thermodynamics renounced understanding the internal structure of the system. It could only, as it were, perceive the surface. But very different organizations can be hidden under the same surface. The information provided by statistical analysis about the system is, therefore, incomplete, and all the more incomplete as the number of internal structures that can be expressed by the same average behaviour is greater. For Maxwell, this information could be obtained gratuitously. Man did not have the sensorial equipment to obtain it; but the demon in the gas-filled tank could, at no cost, estimate the value of the molecules and sort them. For Szilard and Brillouin, on the contrary, information has to be paid for. The demon can 'see' the molecules only if he has with them some physical connection, such as radiation. Not only the gas, but the whole system composed of the gas and the demon tends towards equilibrium. Sooner or later, the demon becomes 'blind' to the gas. He can go on distinguishing the particles only at the expense of energy supplied to the system from without, in the form of light, for instance. In return, the demon obtains the required information about the molecules and, by sorting them, lowers the entropy of the system. But in the long run, total entropy is increased. Even the demon itself cannot escape the all-powerful second law of thermodynamics. The system only works by means of a series of successive transformations involving information. Entropy and information are as closely connected as the two sides of a coin. In any given system, entropy provides a measure of both the disorder and man's ignorance of the internal structure; and information of both the order and man's knowledge. Entropy and information are evaluated in the same way. One is the negative of the other.

This isomorphism of entropy and information establishes a link between the two forms of power: the power to do and the power to direct what is done. In an organized system, whether living or not, the exchanges, not only of matter and energy, but also of information, unite the components. Information, an abstract entity, becomes the point of junction of the different types of order. It is at one and the same time what is measured, what is transmitted and what is transformed. Every interaction between the members of an organization can accordingly be considered as a problem of communication. This applies just as much to a human society as to a living organism or an automatic device. In each of these objects, cybernetics finds a model that can be applied to the others: a society, because language constitutes a typical system of interaction between elements of an integrated whole; an organism, because homeostasis provides an example of all the phenomena working against the general trend towards disorder; an automatic device, because the way its circuits are geared defines the requirements of integration. In the end, any organized system can be analysed by means of two concepts: message and feedback regulation.

Message means a series of symbols taken from a certain repertory – signs, letters, sounds, phonemes, etc. A given message thus represents a particular selection among all the arrangements possible. It is a particular order among all those permitted by the combinative system of symbols. Information measures the freedom of choice, and thus the improbability of the message; but it is unaware of the semantic content. Any material structure can therefore be compared to a message, since the nature and position of its components, atoms or molecules, are the result of a choice made from a series of possible combinations. By isomorphous transformation according to a code, such a structure can be translated into another series of symbols. It can be communicated by a transmitter to any point on the globe where a receiver reconstitutes the message by reverse transformation. This is how radio, television and the secret service work. According to Norbert Wiener, there is no obstacle to using a metaphor 'in which the organism is seen as a message'.[2]

Feedback is a principle of regulation that allows a machine to adjust its activity, not only in terms of what it has to do, but also in terms of what it actually is doing. It operates by introducing into the system the results of its past activity. This brings into action sense organs responsible for estimating the activity of the motor organs, for verifying their performances and making the necessary corrections. This supervision is meant to correct the mechanism's tendency towards disorganization, that is, to reverse temporarily and locally the direction of entropy. These mechanisms range in complexity from the simple regulation of a boiler in relation to the surrounding temperature, to a real system of learning. Every organization calls on feedback loops that keep each component informed of the results of its own operation and consequently adjusts it in the general interest.

With the possibility of carrying out mechanically a series of operations laid down in a programme, the old problem of the relations between animal and machine was posed in new terms. 'Both systems are precisely parallel in their analogous attempts to control entropy through feedback,' said Wiener.[3] Both succeed by disorganizing the external environment, 'by consuming negative entropy', to use the expression of Schrödinger and Brillouin. Both have special equipment, in fact, for collecting at a low energy-level the information coming from the outside world and for transforming it for their own purposes. In both cases, it is the realization, not the intention, that adjusts the action of the system on the outside world through the intermediary of a regulatory centre. An organism preserves a certain stability only through continually borrowing from outside. Despite changes in surroundings, it succeeds in oscillating around its own characteristic equilibrium. It manages to maintain homeostasis because numerous regulatory mechanisms enable it to define the most favourable conditions for its existence. Living or not, every system that functions tends to wear out, to fall into disrepair, to increase in entropy. By means of a certain regulation, each local loss of energy is compensated by work provided by another part of the organism; hence another increase in entropy, in turn compensated by further work carried out at another point in the body. And so on, in a sort of waterfall, by which loss of order in one place is compensated by increased order elsewhere. The coordination of the system depends on a network of regulatory circuits by which the organism is integrated. But as in a waterfall, the total change of energy of all the operations always takes place in the same direction, that imposed by the second law of thermodynamics. The statistical tendency to disorder gradually dilapidates any system that is closed to all exchanges with the outside world. Ultimately, the maintenance of a living system in good repair has to be paid for: the return to the ever unstable equilibrium leads to a deficit of surrounding organization, that is, to an increase in disorder of the total system composed of the organism and its environment. The living organism, therefore, cannot be a closed system. It cannot stop absorbing food, ejecting waste-matter, or being constantly traversed by a current of matter and energy from outside. Without a constant flow of order, the organism disintegrates. Isolated, it dies. Every living being remains in a sense permanently plugged into the general current which carries the universe towards disorder. It is a sort of local and transitory eddy which maintains organization and allows it to reproduce.

Animal and machine, each system then becomes a model for the other. The machine can be described in terms of anatomy and physiology. It has executive organs activated by a source of energy. It has a whole series of sense organs that respond to stimuli from light, sound, touch and heat, in order to watch over its health, sense its environment, verify its nutrition. It contains automatic control centres to estimate its performances; a memory in which the details of the actions to be carried out are prescribed and data of past experience are written down. All this is connected by a nervous system that, on one hand, carries the impressions from the senses to the brain and, on the other, transmits the commands to the limbs. At any time, the machine that executes its programme is capable of directing its action, of correcting or even interrupting it, in accordance with the message received.

And vice versa, an animal can be described in terms of a machine. Organs, cells and molecules are thus united by a communication network. They constantly exchange signals and messages in the form of specific interactions between constituents. The flexibility of behaviour depends on feedback loops; the rigidity of structures on the execution of a programme rigorously laid down. Heredity becomes the transfer of a message repeated from one generation to the next. The programmes of the structures to be produced are recorded in the nucleus of the egg. According to Schrödinger, the chromosomes

contain in some kind of code-script the entire pattern of the individual's future development and of its functioning in the mature state . . . The chromosome structures are at the same time instrumental in bringing about the development they foreshadow. They are law-code and executive power – or, to use another simile, they are architect's plan and builder's craft all in one.[4]

The order of a living organism therefore is based on the structure of a large molecule. For reasons of stability, the organization of a chromosome becomes comparable to that of a crystal. Not the monotonous and rather boring structure in which the same chemical pattern is repeated *ad infinitum* with the same period in all three dimensions; but what physicists call an 'aperiodic crystal' in which the arrangement of several patterns offers the variety that the diversity of living beings requires. A small number of patterns is sufficient, adds Schrödinger. The combination of two signs in the morse code enables any text whatsoever to be coded. The plan of the organism is mapped out by a combinative system of chemical symbols. Heredity functions like the memory of a computer.

Until about the middle of the twentieth century the structure of macromolecules had remained almost completely inaccessible to examination. This situation was changed – once again! – through the convergence of different fields: technology, which found in polymers a new source of material for industry; and physics and chemistry, which tried to purify macromolecules, to determine their composition and to define their organization. The existence of polymers had been recognized in living beings since the beginning of the nineteenth century. On several occasions, indeed, chemists had noted that certain large molecules, when hydrolysed, released only a few simple compounds, sometimes even one. Thus in cellulose or starch, only glucose was found, and in rubber only isoprene. Since Berzelius, the expression 'polymer' had been used to describe the large chemical structure, and 'monomer' the basal sub-unit. These sub-units generally appeared to be linked end to end in the form of a chain. Similar composition, however, did not exclude different structure. Although both are formed from glucose, starch and cellulose nevertheless have very different properties: starch is a food of man, while cellulose is not. Only the arrangement of the sub-units can account for this difference. In some polymers, the sub-units are, in fact, all arranged in the same direction in a uniform chain; in others, in contrast, they sometimes point one way and sometimes the other in alternative patterns. Some chains are long, others short. Some are linear, others branched. In some polymers, only one species of monomer is found, while others contain several. So the variation of a few parameters is enough to generate diversity.

At the beginning of the present century, chemists tried to investigate in the laboratory the means used by nature to construct these huge architectures. The chemistry of polymers is different from that of small compounds and is based on different principles. To prepare a small organic molecule, the chemist has to intervene in each stage and place each atom in the desired place, often at great cost. The production of polymers, on the other hand, requires that the basal sub-units be mixed in certain proportions and exposed to suitable conditions of acidity, temperature, pressure and so on. Once the reaction is initiated, it continues spontaneously without any further intervention by the chemist. He can, however, affect the nature of the final product in a number of ways: by varying the conditions of reaction, by changing the basal sub-units, by modifying their proportions and, especially, by adding to the mixture certain substances that act as catalysts and somehow guide the orientation of the sub-units along the chain. The use of catalysts makes it possible to direct the reaction and to master the organization in space of the product. This led to a close collaboration between science and industry. A whole series of new substances was created by polymerization of simple molecules, particularly hydrocarbons. But a whole series of novel concepts and techniques also emerged and gradually influenced the approach of biology to the macromolecules of living beings.

In this way, the technology of polymers coalesced with the methods of investigation gradually defined by physics and chemistry. The very nature of macromolecules endows them with special properties of weight, electric charge, light diffraction and viscosity. These characteristics provide the means of treating such compounds and studying their behaviour. One can weigh them, so to speak, by subjecting them to a centrifugal force some hundred thousand times greater than gravity. One can measure their mobility in an electric field, estimate their size, volume and general shape. In short, one can form an overall picture of these molecular structures. It was mainly the use of three techniques, however, that revealed the detailed composition of macromolecules, their internal organization and the processes of their synthesis.

The first technique belongs to chemistry. At the beginning of the twentieth century, botanists had found a way of purifying and isolating various plant pigments. They poured plant extracts through long columns of powdered calcium carbonate, which they then washed with various solvents. Retained on the columns in different zones, the pigments separated from each other during elution. This method, known as 'chromatography', was taken up, extended and diversified by chemists in the middle of the present century. It lends itself to infinite variations: one can modify the composition of the column, the solvents used for elution, the activity and concentration of ions, and so on. This technique has great resolving power. It distinguishes between very similar compounds that differ only in details of electric charge, size and shape. Instead of the column, one can even use sheets of special filter paper on which the products to be analysed are run first in one direction, then the other. Each compound moves at its own particular velocity. The chemist can then easily distinguish qualitatively and quantitatively between almost identical compounds. It is no exaggeration to say that the simplicity and efficiency of this technique have entirely transformed the investigation of biological macromolecules, especially proteins and nucleic acids. Previously, these compounds were known to be composed of several chemical sub-units – about twenty amino acids for proteins and four purine and pyrimidine bases for nucleic acid – assembled in very large numbers in each molecule. With considerable effort, it was even possible to hydrolyse a molecule, define the nature of its components and count the numbers of copies of each sub-unit. But there was no way to investigate the arrangement of the sub-units and the spatial organization of a protein. Chromatography made this study possible. By using certain enzymes, the chemist has the means of splitting a protein molecule and collecting, not the separate basal sub-units, but fragments which are themselves formed of several sub-units. He can then cut each fragment into smaller pieces and investigate the composition of each. As in working a jig-saw puzzle, it is then a matter of marking the relative position of the fragments and pieces, and fitting them together to form the original design. Since there are specific enzymes which split the molecule at different points, the same pattern can be cut up into a number of different puzzles. Proceeding step by step, it is possible to obtain enough equations to solve all the unknowns. To everyone's surprise, the extraordinarily complex three-dimensional architecture of the protein molecule turned out to be reducible to a particularly simple structure in one dimension. It is in fact, a linear polymer formed by linking end-to-end some hundreds of sub-units, the twenty amino acids. Three-dimensional complexity results from the folding of the chain on itself, from the twists that confer on the surface an irregular contour. What determines the specific shape of the molecule is the length of the chain, from a hundred to a thousand sub-units long, and the sequence in which the sub-units are arranged. Once again, diversity and complexity spring from the simplicity of a combinative system.

The second technique which transformed the possibilities of biochemical investigation was the discovery of radio-isotopes by physicists. A radioactive element emits radiations that can be detected. It is therefore 'visible' wherever it occurs in an organism. A chemist can place a radioactive atom by synthesis at a chosen point in an organic or inorganic molecule. Once it has entered the organism, such a molecule can be traced. It is possible to observe its successive transformations, its distribution among the organism's constituents and its retention or excretion. The use of isotopes made it possible to unravel the tangled threads of intermediary metabolism, to follow step by step the elaboration of small molecules and their polymerization into large ones, to measure the stability of chemical constituents or the rate at which they are renewed. When coupled with histological examination, autoradiography of radio-isotopes even provides a means of detecting the constitution of cell structures visible by microscope and watching their changes during the cell-cycle. It was no longer only the composition of living beings that was an object of investigation; it was all the dynamics of their chemical transformations.

Finally, the third technique that provided access to the structure of the macromolecules in the cell came from improvements by physicists of the means of observation. First, with the electron microscope, in which the substitution of a beam of electrons for visible light increases resolving power more than a thousand fold: it then became possible to observe the fine details of cell organelles and even to distinguish the shape of certain very large molecules. Secondly, and more particularly, through study of crystals by X-ray diffraction: in addition to the general form of a molecule, the exact position of each atom could be perceived, not only in simple compounds, but even in the largest molecules. To the precision of results, however, there corresponds the difficulty of the technique. Reserved for the physicists, it led some of them to become directly interested in biology. X-ray diffraction had been used in Great Britain at the beginning of the twentieth century to analyse the organization of simple crystals such as sodium chloride. From that was born a school of crystallographers interested in resolving the internal structure of all kinds of compounds; even of biological macromolecules, so convinced were physicists of this school that the functions of the living cell must be based only on the configuration of these molecules. It was actually one of them who suggested the expression 'molecular biology' to describe this type of analysis. Every specialist occupied in some way with experimentation on the cell then found himself in the position of Monsieur Jourdain, doing molecular biology 'without knowing it'. Somewhat isolated at first, crystallographers groped their way through the complexity of biological systems. They had to proceed by stages from the simple to the complex, gradually increasing the power of resolution, learning to recognize certain privileged areas in molecules and marking these with heavy atoms easy to detect with X-rays. Little by little, they became able to discern the contours of the macromolecules and even to define the details. Purely crystallographic analysis was then replaced by an elaborate game combining the interpretation of physical data, the construction of laboratory models and a kind of intuition based on knowledge of atomic properties and bondings. However arduous it might be, this new-born collaboration between crystallography and theoretical chemistry proved to be the only way of investigating the organization of large biological molecules. For although chemical analysis describes the way the sub-units are arranged along the chain, it provides no information about the folds of the chain, the detailed anatomy of the molecule or its configuration in space. Physical analysis makes it possible to localize every detail in a molecule containing several thousand atoms.

The crystallographers were joined by another contingent of physicists, also interested in biology but for different reasons. After the Second World War, many young physicists were shocked by the military use made of atomic energy. Some of them, moreover, were dissatisfied with the direction taken by experiments in nuclear physics, by their slowness and by the complexity inherent in the use of large machines. They saw in this the end of a science, and looked around for other activities. Some turned to biology with a mixture of anxiety and hope: anxiety, because all they usually knew about living organisms consisted of vague recollections of zoology and botany acquired at school; hope because some of their most celebrated elders pointed to biology as a science full of promise. Niels Bohr saw it as the source of new laws of physics awaiting discovery. Schrödinger also prophesied a new and exhilarating era for biology, particularly in the field of heredity. Just to hear one of the leaders in quantum mechanics asking 'What is life?' and then describing heredity in terms of molecular structures, inter-atomic bonds and thermodynamic stability was enough to fire the enthusiasm of certain young physicists and to bestow some sort of legitimacy on biology. Their ambition and interest were limited to a single problem: the physical basis of genetic information.

Micro-organisms

Not only the concepts and techniques, but also the material used in the study of the cell and heredity changed in this mid century. Classical genetics had not been able to bridge the gap between gene and character. It had come to the conclusion that the chromosomes contain a substance capable both of reproducing itself exactly and of carrying genetic specificity. But the biological objects used in genetics during the first part of the twentieth century lent themselves neither to a search for this kind of substance nor to the analysis of its mode of action. Some efforts to combine physiology and genetics in the study of *Drosophila* had succeeded in demonstrating the influence exercised by genes on certain chemical reactions of the organism. But in a complex organism which reproduces sexually, the effect of a gene is usually expressed only after a long lapse of time and after a series of transformations imposed by development and morphogenesis. The organisms studied by geneticists did not suit chemists, and vice versa. In order to combine their efforts successfully, they had to find a common experimental material. Contrary to all expectations, this material was provided by micro-organisms, especially bacteria and viruses.

Born almost three centuries ago with the invention of the microscope, bacteriology had long remained observational, until transformed into an experimental science by the work of Pasteur. Within a few years, man was astonished to discover that, without microorganisms, the world would not be what it is. However, the importance of microbes as pathogenic agents, their function in the cycles of matter in the biosphere and their role in many industries had long overshadowed their value in the study of biological mechanisms.

The cell theory had contributed to the unification of living organisms; but bacteria had remained excluded from the cellular world. Their small size, in fact, prevented the recognition of the characteristic structures. It was scarcely possible to do more than cultivate them, describe them, and attempt to classify them. Only at the beginning of the present century did microbes gradually become objects of investigation by physiologists and biochemists. The development of medicine and industry required accurate identification of germs and knowledge of their properties. The greater the number of microbes isolated, the more important it became to define and distinguish them. Although microbiologists confined themselves to studying growth under different conditions, they nevertheless succeeded in defining the nutritional requirements of micro-organisms, their ability to use certain compounds as a source of carbon and their sensitivity to antimicrobial substances. At the same time, chemists found micro-organisms to be especially suitable for their studies. Simpler to handle and more reproducible than pigeon-muscle or rat-liver, a culture of yeast or bacteria was just as suitable for the extraction of constituents, the investigation of metabolism or the determination of enzymatic activities. Pigeons, rats and bacteria all exhibited remarkably similar properties: they all showed the same chemical reactions, the same intermediates with high-energy potential, the same enzymatic activities always associated with proteins. Behind the variety of shapes and the diversity of properties, there appeared a unity of composition and function throughout the living world. It was always the same materials that were used to produce the same constituents; as if most often, nature had only one way of operating.

Until about the middle of the twentieth century, however, there appeared to be nothing in common between micro-organisms and higher organisms in one area: heredity. Only lately recognized as a unit of mutation and function, the gene was chiefly considered as the unit of recombination and segregation. Genetics was based on the study of sexually reproducing hybrids. The role of chromosomes and the mechanics of heredity had been established by a combination of genetic investigation and cytological observation. But nothing of that kind could be achieved for microbes, neither by hybridization nor by cytology. They reproduced vegetatively and showed no sign of sexuality. Their small size precluded cytological observations. Their lack of organization made it impossible to distinguish between somatic and germinal elements, between character and factor, between phenotype and genotype. Both bacteriologists and geneticists agreed, therefore, that bacteria were devoid of a genetic apparatus, and that their heredity had nothing in common with that of animals and plants. The world of microbes seemed not to obey the concepts and the methods of genetics.

Only in the middle of the present century did micro-organisms become accessible to genetic investigation. First moulds and yeasts, where phenomena of sexuality and conjugation were observed. In these organisms, the study of metabolism and of heredity could be combined. The characters they displayed were no longer secondary properties, such as the length of a wing or the colour of a flower, but were expressed in the very chemistry of the organism, its power of growth, its synthetic ability. For the first time, a geneticist and a chemist joined forces to study a mould that multiplies and forms all its constituents in a simple medium. The geneticist isolated mutants that had become unable to grow in this medium. The biochemist sought the reason for this inability. It then appeared that each mutation blocked metabolism at a certain point; each prevented the synthesis of an essential metabolite; each impaired the quality of an enzyme involved in this synthesis. The whole chemistry of the organism was therefore governed by its heredity. A particular gene controlled a particular chemical reaction because it determined the properties of the particular protein-enzyme that catalysed the reaction. So protein made its appearance in the gap previously observed between gene and character.

Once the metabolic reactions had been thus transformed into objects for genetic study, it became possible to analyse the heredity of such simple organisms as bacteria. Those very properties which had heretofore excluded bacteria from genetics then made them particularly suitable material for the study of variation. Small in size and rapid in growth, bacteria produce, in a few hours, enormous populations in very small volumes. By applying statistical methods to these cultures, variation was found to be the result of rare quantum changes, identical with mutations in higher organisms. Like flies, bacteria therefore had hereditary factorial determinants, that is, genes. The genes governed morphology as well as metabolism and all discernible properties. In certain microbes, there were even phenomena of conjugation recalling sexuality in higher organisms, with male and female bacteria. It then became possible to produce hybrids and to define the relations between genes. These turned out to be arranged along linear structures similar to the chromosomes in higher organisms. The same applied to viruses. In the entire living world, therefore, there is but one way of ensuring permanence of forms and properties through succeeding generations. There is also but one way of modifying them. The rules of the genetic game are the same for all.

Previously, sexual reproduction appeared to be the only means of reassorting the genes of a species in combinations sufficiently numerous to allow an almost infinite variety of individuals. But in bacteria there are other means than sex for transferring genetic material from one cell to another. Thus viruses can act as vehicles for bacterial genes and produce a kind of infectious heredity. Also certain species can absorb and incorporate into their own chromosome the genes released by crushing other bacteria. It was wrong, therefore, to establish a correlation between heredity and sexual reproduction. The notion of heredity had to be extended. It became the ability to reproduce its like possessed by every cell and transmitted through successive generations. The central principle of heredity is this ability to reproduce structures and reactions through multiplication. No living organism exists without it. All the rest, sexuality, variety of shape and differentiation of cells, are just complications elaborated in the course of evolution, just variations on the same fundamental theme. It is perfectly possible to imagine a rather boring universe without sex, without hormones and without nervous systems; a universe peopled only by identical cells reproducing *ad infinitum*. This universe, in fact, exists. It is the one formed by a culture of bacteria.

The use of bacterial cultures as experimental objects had two important consequences. First, it provided access to genetic fine structure. In fact, genetic investigation of bacteria was simplified to the extreme and acquired a power of resolution unknown in the study of complex organisms. The simple gesture of spreading a few drops of culture on selective media provides, within a few hours, information about billions of events, mutations and recombinations. It is difficult to imagine how much work would be required to obtain an equivalent result by the study of animals or plants. This increased power of resolution led to revising the classical picture of the gene as an integral structure, a bead on a necklace. The gene, though defined as a unity of function, in fact contains several hundred elements that mutation can modify and recombination separate. But to this new organization the old principles still apply: the quantum nature of variations, the role of chance in the reassortments subject to the laws of probability, the arrangement of the elements along a one-dimensional structure. Moreover, bacteria provided access to the chemistry of genetic material. If genes released by crushing of bacteria could really penetrate other bacteria, if they could take root in them and give them new characters, then the chemist could intervene. He could extract genes, measure and purify them like any other compound. That is how Avery found genetic activity to be linked to the presence of deoxyribonucleic acid. For almost a century, this substance had been known to be present in the cell nucleus; its overall chemical composition was known. But until that time, there had been no way to give it a role or a molecular structure. Its role is to carry the specificity of Mendel's units. Its structure can be determined by a combination of chemical analysis and crystallography. It is a long polymer formed by the alignment of four sub-units, the four organic bases, repeated by millions and permuted along the chain, like the letters of the alphabet in a text. It is the order of these four sub-units that directs the order of the twenty sub-units in proteins. Everything then leads one to regard the sequence contained in genetic material as a series of instructions specifying molecular structures, and hence the properties of the cell; to consider the plan of an organism as a message transmitted from generation to generation; to see the combinative system of the four chemical radicals as a system of numeration to the base four. In short, everything urges one to compare the logic of heredity to that of a computer. Rarely has a model suggested by a particular epoch proved to be more faithful.

The use of bacterial cultures by different disciplines had a second consequence: with so rudimentary an organism, different techniques could be simplified to the point that they could be applied simultaneously. To investigate heredity, it was therefore no longer sufficient to observe the characters of the bacterial cell, their variations and reassortments in hybrids. It was necessary, in parallel, to extract the genetic material, to specify its characteristics, and to measure its properties in a centrifuge. At the same time, it was also necessary to analyse the corresponding proteins, to define their structure and to estimate their enzymic activity. In that way, it was possible to follow the effects of a mutation, not only through the changes in function, but also through the changes in the structures involved. Conversely, it became possible to analyse the properties, organization and function of a cell through the study of the lesions caused by mutations. Genetic analysis, then, no longer simply aims at dismantling the mechanism of heredity: it becomes a precision instrument for detecting the constituents of the cell, and their role and interactions with other elements. Mutations are used to dissect the cell without destroying it. For more than a century, the study of pathology had provided one of the surest methods for interpreting the normal state. Experimental physiology intervened from outside the organism, producing lesions by mechanical means or by the effect of toxic substances. Molecular biology provokes lesions from inside the organism as a result of mutations. It seeks to affect, not the structures already formed, but the programme that directs their formation. The selective pressure exerted on a bacterial population in the laboratory can become so great that it is possible to obtain almost at will monsters in which a chosen function is damaged by mutation. By means of such monsters, abnormalities are analysed by physiologists and morphologists in the intact organism, by biochemists and physicists in cell-free extracts. There are no longer two independent fields, but two aspects of the same investigation.

What changes had been wrought in the attitude of biology! Its whole history had been dominated by a conflict between two opposite approaches: between the hope of explaining the performances of organisms in terms of the properties of matter, and the refusal of a conception in which the integration of living beings is denied any particular property. On the one hand, the more chemical investigation advanced, the more it showed the identity of the laws governing the living and inanimate worlds. On the other, the more the study of the behaviour and evolution of living organisms was extended, the more it showed breaks at each level of integration. From virus to man, from cell to species, biology is interested in systems whose complexity is constantly increased by the successive integration of lower-level systems. Each level of organization represents a threshold where objects, methods and conditions of observation suddenly change. Phenomena that are recognizable at one level disappear at the lower level; their interpretation is no longer valid at a higher level.

Biology has then to articulate these levels two by two, to cross each threshold and unveil its peculiarities of integration and logic. Bacteria represent a vital minimum, so to speak. Ignoring the other organisms, molecular biology began at the very lowest level of integration characterizing an organism with bacteria. It deliberately established itself at one of the frontiers of the living world, at the limit of the inanimate. The level below is described in terms of chemistry and physics, the level above in terms of organization, of logical system or even of an automated machine. But the nature of the system and the methods of observation always make it possible to consider the two levels together, constantly to compare the whole with the details of constituents or phenomena revealed by analysis. Our knowledge of the bacterial cell has developed through the accumulation of various techniques practised either on the whole organism, or on extracts; in particular, through the combination of genetic investigation with physical and chemical analysis at each stage. It is necessary to destroy the integrity of the cell in order to recognize the components and study their function. But it is necessary to use intact cells to verify that, after isolation and purification, isolated compounds really act in the test-tube as they do in the organism. It was from the comparison of these two levels that the description of the bacterial cell emerged.

The Message

An individual bacterium is rarely an object of investigation. Experimentation generally involves a culture, a medium containing a bacterial population that readily attains billions of microbes in a few cubic centimetres. There is thus no hope of directly observing the properties of each individual. Even to detect the presence of certain particular individuals, of mutants, for instance, it is necessary to place a population under such conditions that only these individuals can multiply. In that case it is not the individuals themselves that are observed, but their descendants – again an enormous population. The image that the biologist forms of the bacterial cell can therefore be only statistical. It is a sort of composite picture that emerges from a mass of observations collected from a mass of individuals.

This composite picture represents the simplest object that combines the properties of organization, autonomy and invariance generally considered to characterize living beings. More rudimentary organizations, such as viruses, certainly exist; but although viruses exhibit certain properties of organisms, they are far from possessing them all. The consequent lack of autonomy prevents their being considered as living organisms. As for the cells found in higher organisms, or in unicellular organisms such as protozoa and yeasts, they have far more complex organization than a bacterium.

The relative simplicity of the bacterial cell, however, does not mean that it can be considered as primitive from the evolutionary point of view. It is tempting to link simplicity with archaism and consider the bacterial cell as a living fossil, or even as our common ancestor. But whether bacterium or mammal, every living organism that the biologist can examine is the result of billions of years of evolution. When the paths followed by evolution no longer have visible landmarks, their meanders are likely to remain forever unknown. Nothing enables us to measure the kinship between primitive organisms and existing bacteria. The original material used by evolution can only be imagined. Even if it bore some resemblance to today's bacterial cell, we cannot conceive of it as having had the same disconcerting complexity. Behind each bacterium that we examine, behind each of its components, there is a long history as necessary to an understanding of the system as the knowledge of its structure.

For technical reasons, study of the bacterial cell has concentrated on a harmless bacterium, *Escherichia coli*, an organism normally present in the human intestine. It multiplies perfectly well in a medium composed of a few simple mineral salts and an organic compound – sugar, for instance – that provides it with both carbon and energy; there is nothing in this medium that the chemist cannot produce from simple ingredients. The bacterium *Escherichia coli* can also be cultivated in a more complex medium, such as meat broth, which contains various organic products. The presence of these organic compounds, necessary for growth, allows the bacterial cell to skip a number of syntheses and thus to multiply more quickly. Under optimum conditions of temperature and aeration, this bacterium succeeds in reproducing every twenty minutes. It grows and lengthens during twenty minutes. Then it cleaves into two, producing two bacteria identical with each other and with the original one.

A bacterium is far too small an object to be visible to the naked eye. Under a light microscope, *Escherichia coli* appears elongated in shape, about one micron long and half a micron wide; but no internal structure can be detected. Under an electron microscope, photographs of sections of the bacilli, previously fixed and stained, recall lunar landscapes. Only by perseverance and comparison, using a combination of numerous techniques, have a few visible structures been identified. In the electron micrographs, the bacterial cell appears as a small bag which keeps its shape by means of a rigid wall. Beneath this wall, the bag is surrounded by a two-layer membrane that forms a physical separation between cell and medium. Impermeable to the passage of certain substances, the membrane protects the cell against loss of the molecules it produces, but allows certain mineral salts to circulate freely. In addition, by means of some kind of pumps in the membrane, the bacterial cell can absorb and concentrate certain compounds, such as sugars, that it finds in the medium and that it requires for its metabolism. Little is known as yet about the structure of the membrane and the way the pumps work. Even under the electron microscope, only a small number of defined structures can be distinguished within the bag. In the central area, there appears to be a fairly dense mass formed of fibres folded on themselves and twisted into skeins: this is the long fibre containing the genetic programme. Apart from this, the bag seems to contain only thousands of small spherical granules all the same size: this is where the proteins are synthesized.

When he opens the bag, the chemist finds several thousand molecular species. These molecules, however, do not form a continuous series in size. They can be classified in two well defined categories. About half of them are quite small, not exceeding molecular weights of 500 to 600. The others, in contrast, are very large molecules, greater than 10,000 to 20,000. There is no intermediate size between the two. Astonishing at first sight, this distribution is explained by the way the cell elaborates its constituents. To construct enormous molecular architectures piece by piece, atom by atom, represents a very complex and costly task, even for the cell. The cell, therefore, operates in two stages. First, the elements taken from the medium are combined through a series of transformations. Either one by one or in groups, they are constantly exchanged, moved about, added or subtracted. This first step essentially consists in linking the carbon atoms together. It gives rise to the skeletons of different structures, elongated or closed in rings, on which other atoms are later hooked. All this activity involves hundreds of chemical reactions. But ultimately, it produces a limited number of small compounds, a few dozens at the most. In the second stage of cellular chemistry, the small molecules are assembled to produce the larger ones. It is the polymerization of sub-units linked end-to-end that forms the characteristic chains of macromolecules. Each chemical operation is, then, exactly the same as the others; it always involves the addition of one sub-unit to a growing chain. By varying the length of the chain, by arranging the sub-units in a different order, the cell thus builds up a considerable number of large structures from a limited number of simple sub-units. The two stages of cellular chemistry, therefore, differ in their function, products and nature. The first stamps out the chemical patterns; the second assembles them. The first forms compounds that exist only temporarily, for they are intermediate stages in biosynthesis; the second constructs durable products. The first operates by a series of different reactions; the second by repeating the same reaction.

A series of transformations involving small compounds is also used by the bacterial cell to transform the energy it extracts from the medium. In the world of bacteria, there are many ways of obtaining energy: by capturing solar radiations or by oxidizing organic or inorganic compounds. But to accumulate and transfer its reserves, the bacterial cell again functions like all living organisms. The same phosphorus-rich compound is always involved: its synthesis allows the potential energy to be stored; its hydrolysis allows the energy to be mobilized when required. Our bacterium *Escherichia coli* can extract its energy only from the decomposition of certain organic compounds, such as sugars. In a sugar molecule – glucose, for instance – the atoms are arranged in a well defined structure according to a precise spatial order. By destroying the molecule and disorganizing its structure, the bacterial cell converts the original order in the glucose into chemical energy. This energy is then used to synthesize bacterial constituents – in other words, to establish a different molecular order, that of the cell. Ultimately, therefore, mobilization of energy is expressed by transfer of organization, by conversion of the order of the medium into the order of the bacterium. To decompose a sugar molecule, the cell proceeds progressively through a series of reactions. At certain stages, a quantum of energy is released and collected as a molecule of this phosphorus-rich compound. Thanks to this controlled breakdown, the transfers of energy take place in the cell with a high yield.

If analogy is to be used, the bacterial cell is obviously best described by the model of a miniaturized chemical factory. Factory and bacterium only function by means of energy received from the exterior. Both transform the raw material taken from the medium by a series of operations into finished products. Both excrete waste products into their surroundings. But the very idea of a factory implies a purpose, a direction, a will to produce – in other words, an aim for which the structure is arranged and the activities are coordinated. What, then, could be the aim of the bacterium? What does it want to produce that justifies its existence, determines its organization and underlies its work? There is apparently only one answer to this question. A bacterium continually strives to produce two bacteria. This seems to be its one project, its sole ambition. The little bacterial cell performs at top speed the two thousand or so reactions which constitute its metabolism. It grows. It gradually elongates. And when the time is ripe, it divides. Where there was one individual, suddenly there are two. Each of these individuals then becomes the centre of all the chemical reactions. Each manufactures all its molecular structures. Each grows anew. A few minutes later, each divides in turn to produce two individuals. And so on, for as long as conditions permit. For two billion years or more, bacteria – or something like them – have been reproducing. Structure, function and chemistry of the bacterial cell, all have been refined for this end: to produce two replicas of itself, as well as possible, as quickly as possible and under the most varied of circumstances. If the bacterial cell is to be considered as a factory, it must be a factory of a special kind. The products of human technology are totally different from the machines that produce them, and therefore totally different from the factory itself. The bacterial cell, on the other hand, makes its own constituents; the ultimate product is identical with itself. The factory produces; the cell reproduces.

The two kinds of synthesis carried out by the living cell – successive rearrangements and polymerization – are not fundamentally different from those carried out in the laboratory by the organic chemist. There is no particular mystery about transformations that occur in the cell; no unknown material; no reaction or chemical bond that appears beyond the reach of laboratory techniques. Not only can the chemist prepare many of the compounds found in the cell; some of them even arise spontaneously under conditions that, in all likelihood, obtained on the earth's surface before the appearance of living organisms. This happens, for instance, when solutions of appropriate inorganic composition are 'excited' by discharges of energy, ultra-violet radiation, for instance. Neither in the raw materials, nor in the nature of the reactions, nor in the type of bonds formed does any discontinuity appear between the chemistry of the living and that of inanimate matter.

But although laboratory and industry are able to produce some of the compounds characteristic of the cell, at what tremendous cost! The apparatus is expensive and cumbersome; the yields lamentably small; the conditions of temperature, pressure and acidity almost always incompatible with life. And during this time, our bacterial cell carries out some two thousand distinct reactions with incomparable skill, in the smallest space imaginable. These two thousand reactions diverge and converge at top speed, without ever becoming tangled, and produce exactly the quantity and quality of molecular species required for growth and reproduction, with a yield close to one hundred per cent. The chemistry of living organisms is different from that of the laboratory not in the nature of the work accomplished, but in the conditions under which it is performed.

And yet, the secret of the cell has long been known to chemists: it is the use of catalysts, those substances which activate a reaction without taking part in it or being themselves chemically transformed. Most chemical reactions occur spontaneously but extremely slowly under laboratory conditions or in the cell. The catalyst only increases the rate of the reaction. But whereas laboratory catalysts are generally unspecific, those of living organisms are strictly specific. For each chemical reaction in a cell, there exists one particular catalyst, one particular enzyme, and one only. Each catalyst activates one of the cellular reactions and only that one. To perform its two thousand chemical operations, the cell therefore has to produce two thousand different kinds of enzymes. All enzymes belong to the same molecular family, proteins. Not all proteins are enzymes: some play another role in the cell. But all enzymes are proteins. Each contains several thousand atoms assembled in strict order. The geometry of its structure gives an enzyme its properties. Changing one single chemical radical, transferring a few atoms may be sufficient to deform the molecule and make it lose its function.

All the chemistry of the cell, its precision and efficiency are therefore based on the properties of some two thousand protein-enzymes that catalyse metabolic reactions. What is reproduced exactly at each generation is, therefore, simply the bacterial cell as a whole. It is each of the enzymes that control its chemistry, each of the molecular species that constitute it. The extraction and study of a protein requires a culture containing at least a million million bacteria. One single bacterium contains several thousand molecular species. Yet when the chemist attempts to isolate and purify a protein from this kind of mixture, he succeeds. When he endeavours to analyse the composition of protein and determine the sequence of sub-units in the chain, he succeeds again. When a crystallographer seeks to specify the organization of the molecule and detect the exact position of each atom, he also succeeds. This means that the thousands of copies of the protein, each synthesized by the million million bacteria, all have exactly the same properties; that all are built of exactly the same sub-units placed in the same sequence; that all have exactly the same structure and the same atoms distributed in the same manner. In short, all bacteria of the culture produce strictly the same molecular species. If there are errors, they are too few to be detected.

The permanence of living organisms through successive generations is therefore observed not only in their shapes, but even in the fine chemical details of the substances that compose them. Each chemical species is reproduced exactly from one generation to another. Each chemical species does not, however, form copies of itself. A protein is not born of an identical protein. Proteins do not reproduce. They are organized from another substance, deoxyribonucleic acid, the constituent of chromosomes. This compound is the only one in the cell that can be reproduced by copying itself. This is a consequence of its unique structure. Deoxyribonucleic acid is, in fact, a long polymer formed not of one, but of two chains, helically twisted around each other. Each chain contains a skeleton formed of alternating sugar and phosphate groups. Each sugar molecule is linked to only one chemical residue – an organic base – of which there are four different kinds. These four sub-units are repeated by millions in infinitely varied combinations and permutations along the chain. By analogy, this linear sequence is often compared to the arrangement of the letters of the alphabet in a text. Whether in a book or a chromosome, the specificity comes from the order in which the sub-units, letters or organic bases, are arranged. But what gives this polymer a unique role in reproduction is the nature of the relations that unite the two chains. Each organic base in one chain is associated with one in the other, but not just any one. The system of chemical bonding is such that each sub-unit on one chain can correspond to only one of the other three sub-units in the second chain. If the four sub-units are indicated by A, B, C and D, A in one chain is always opposite B in the other, and D is always opposite C. The symbols go in pairs; the two chains are complementary. The sequence in one chain imposes the sequence in the other.

Owing to these peculiarities of structure, the nucleic acid duplex is exactly reproduced. Since the two chains are complementary, each contains the details of the sequence. The reproduction of the molecule then results from the separation of the two chains, followed by the reconstitution by each chain of the one complementary to it. Since opposite B, there can only be A, and so on, each chain can direct the synthesis of the complementary sequence without ambiguity. Remarkable in its simplicity, this mechanism produces two molecules identical to the original one. Reproducing the duplex of the chromosome really amounts to recopying it sign by sign. The forces responsible for the recognition and placing of each sub-unit are those which control the formation of crystals. To connect each radical with the preceding one, to bring about the chemistry of polymerization, a few enzymes are sufficient. Little is known as yet about these enzymes. Some of them, however, have been isolated. They copy deoxyribonucleic acid in a test-tube when provided with the necessary ingredients for the reaction, that is to say, the four organic bases in a suitable form. In the bacterial cell, repair systems have also been detected, which 'try out' the copies, as it were, check their accuracy and correct certain errors.

Since it is the only bacterial constituent to be re-copied in this fashion, the nucleic acid duplex perpetuates the structure of the other chemical species through successive generations. Its role is, in fact, to direct the synthesis of proteins and guide their organization. A particular gene corresponds to a particular segment of the nucleic acid chain. There the necessary instructions for building a particular protein are coded in a numbering system to base four. Nucleic acid and protein are both linear polymers. Each is characterized by the sequence of sub-units it contains, by the order in which they are arranged along the chain. The nucleic-acid sequence determines the order of the protein sub-units. This is a unidirectional process: the transfer of information always goes from nucleic acid to protein, never in the other direction. But whereas the combinative system of nucleic acid uses only four chemical symbols, that of protein uses twenty. The activity of the gene, the execution of instructions for protein synthesis therefore requires the univocal transformation of one system of symbols to the other.

The representation of the genes envisaged by classical genetics as individual structures arranged like a string of beads has, therefore, been replaced by that of a linear sequence of chemical symbols, the aperiodic crystal predicted by physicists. The model that best describes our knowledge of heredity is indeed that of a chemical message. Not a message written in ideograms like Chinese, but with an alphabet like that of the morse code. Just as a sentence represents a segment of text, so a gene corresponds to a segment of nucleic acid. In both cases, an isolated symbol means nothing; only a combination of symbols has any 'sense'. In both cases, a given sequence, sentence or gene, begins and ends with special 'punctuation' marks. The transformation of a nucleic-acid sequence into a protein sequence is like the translation of a message received in morse that does not make sense until it is translated, into English, for example. This is done by means of a 'code' that provides the equivalence of signs between the two 'alphabets'.

The activity of the genes, the ordering of the sub-units in the protein chains, therefore represent a far more subtle operation than their reproduction, the ordering of the nucleic-acid sub-units. To translate and form the chemical bonds in protein, the bacterial cell deploys a piece of extremely complex equipment. The synthesis of proteins is a two-stage process, since the protein sub-units are assembled and polymerized, not directly on the gene, but on small particles in the cytoplasm which serve as assembly lines. The deoxy-ribonucleic acid text of the gene is therefore first transcribed into another species of nucleic acid, the so-called ribonucleic acid, by means of the same four-sign alphabet. This copy, called the 'messenger', associates with the particles in the cytoplasm and brings them the instructions for assembling the protein sub-units in the order dictated by the nucleic-acid sequence. The translation of the genetic text re-copied in the message takes place through the intervention of other molecules called 'adapters'. These adapters bring the appropriate protein sub-units into juxtaposition with the nucleic-acid sub-units and thus establish a univocal correspondence be-tween the two alphabets. Carrying suitable adapters, the particles move from end to end of the messenger nucleic acid, like the reading head of a tape-recorder passing over the tape. The protein sub-units are thus aligned in the order prescribed by the gene. Each sub-unit is successively attached to the preceding one by an identical chemical bond. The protein chain is thus synthesized step-wise, from one end to the other.

The genetic code is almost completely known today. Each protein sub-unit corresponds to a particular sequence of three nucleic-acid sub-units, of three organic bases, called a triplet. The four nucleic-acid sub-units can give rise to sixty-four different triplets: the cell contains, therefore, a 'dictionary' of sixty-four genetic terms. Three of the triplets provide the punctuation marks; in the nucleic-acid text, they indicate the beginning and the end of the sequences corresponding to the protein chains. Each of the other nucleic triplets 'means' a protein sub-unit, an amino acid. Since there are only twenty protein sub-units, each of them corresponds to several triplets or synonyms in the nucleic-acid dictionary, which provides a certain flexibility in the writing of heredity. All organisms, from man to bacteria, seem able to interpret any genetic message correctly. The genetic code seems to be universal and its key known to the whole living world.

Once the protein sub-units have been arranged in order and linked up, the chain folds back on itself in a complicated and unique pattern. The protein thus acquires its final form, which confers on it particular catalytic – or other – properties. This transformation of a one-dimensional structure into a three-dimensional one is still not understood in detail. It does not seem to require any particular factor not already involved in synthesis. It appears to occur spontaneously, by the simple interaction between the chemical residues distributed along the chain, some of which are mutually attractive, and others mutually repulsive; the sequence, once formed, can convert itself freely into a certain spatial arrangement, into a strictly defined configuration. Proteins can be denatured by certain treatments, physical or chemical. The molecule then loses its shape; it unrolls and becomes an extended chain again. When placed back under physiological conditions, some chains resume their specific three-dimensional structure; others do not. In all likelihood, proteins of the latter type acquire their configuration only during synthesis, by the formation of a 'nucleus' around which the rest of the molecule is organized. In all cases, however, it is only by physical interactions that this key stage in the reproduction of molecules is reached: the transformation of the blueprint into the building, the conversion of potential into functional.

The bacterial cell contains a single molecule of deoxyribonucleic acid, a long duplex in which some ten million signs are arranged. It is more than a millimetre in length, about a thousand times greater than the diameter of the cell in which it is coiled into a skein. During bacterial growth, the duplex is reproduced once only each generation, and each of the two bacteria formed by cell division receives a copy. This molecule of nucleic acid forms the 'chromosome' of the bacterium: it contains all the genes necessary for determining the organization and function of the bacterial cell. The system is therefore arranged with but one end in view: to enable each bacterium to produce two bacteria, each of which will contain a copy of the chromosome and in turn will produce two bacteria.

What is coded along the nucleic-acid chain, what is re-copied sign by sign to be scrupulously transmitted from one generation to the next, is the set of structural plans necessary for the bacterial cell – the ensemble of instructions prescribing in minute detail the whole series of protein structures. The programme is not transcribed and translated at one stroke, but in segments. Reading the message may be compared, not to unrolling a scroll of paper from end to end, but rather to consulting the pages of an instruction book when required. Some parts of the programme contain directions that refer back to others, according to circumstances. They say what has to be done in a given situation. For example, on page 35 there are the following instructions: 'Make an apparatus for detecting the sugar galactose in the culture medium; if galactose is present, follow the instructions on page 341; if not turn over.' Or again, page 428 gives the plan for constructing an apparatus to measure the concentration in the cyto-plasm of an essential metabolite, a protein sub-unit, arginine: 'If this concentration exceeds a certain value, do nothing; if it is below this value, follow the instructions on pages 19, 64, 155, 601 and 883.' Most of the situations that a bacterium might encounter are provided for in the message. The programme thus contains the plans of all the parts required to make a bacterium and gives it the means of facing up to the difficulties of everyday life. But it is only a programme. In the processes that lead to re-copying the nucleic-acid sequence, whether for reproduction or protein synthesis, deoxyribonucleic acid plays the passive role of a matrix. Outside the cell, without the means to carry out the plans, without the apparatus necessary for copying or translating, it remains inert, like a tape outside the tape-recorder. No more than the memory of a computer can the memory of heredity act in isolation. Able to function only within the cell, the genetic message can do nothing by itself. It can only guide what is being done. To produce machines from plans, there have to be machines. None of the substances that can be extracted from the cell can re-produce. Only the bacterium, the intact cell, can grow and reproduce, because only the cell possesses both the programme and the directions for use, the plans and the means of carrying them out.

Regulation

The means of carrying out the plans are the proteins. On their properties all the activities, architecture and integration of the cell are based. Proteins act, not by forming chemical bonds, but by associating with other compounds. Their structure, in fact, gives them a unique quality: the ability to 'recognize' exactly one or more chemical species in the most heterogeneous mixture. The precision and specificity of this choice determine the relations between the constituents of the cell. They govern its whole chemistry.

An enzyme can catalyse the transformation of a particular metabolite, and only that one, because in the irregularities of its surface there exists a kind of crevice into which only this one molecular species fits exactly. Once the substrate is lodged in its site, some of its atoms are subject to action from the surrounding residues of the protein. The forces uniting some atoms of the substrate are perturbed; hence a change in a bond, for instance a break. The product thus formed no longer exactly fits into the protein; it quits the crevice leaving the protein intact and the place free for another molecule of substrate. All this happens in a fraction of a second. The secret of this chemistry lies, therefore, in the precision of the chemical patterns, in the way each enzyme fits its own particular substrate: on these depend the efficiency and rate of the reactions. Since a metabolite is recognized by only one enzyme, it is inevitably directed along a given chemical pathway prescribed by the catalytic activity of the enzyme. In return, the enzyme is fully 'informed' about the nature of one chemical species and ignores others. Its very structure determines both its choice and its interaction. An enzyme thus possesses the very properties that Clerk Maxwell attributed to his demon: in the mixture of compounds contained in the cell, it 'sees' one molecular species, and one only, for which it opens the door to reaction.[5] The strictness of the order characterizing exchanges of matter and energy in the cell depends on the exactitude of the sorting operation performed by the enzyme.

The specificity of the interactions established by some proteins, either with each other or with other compounds, also determines the architecture of the cell. Despite its relative simplicity, the bacterial cell already contains so many different chemical species, it already reaches a degree of complexity so great that it is difficult to perceive its arrangement as a whole. But we are beginning to understand what determines the form of simpler objects, such as viruses. A virus is a particle formed by a fragment of nucleic acid enclosed in a shell of protein. Lacking enzymes and all the chemical equipment necessary for syntheses and for the mobilization of energy, the virus cannot reproduce by itself, but only within the cell that it infects, and whose mechanisms it uses for its own purposes. Only in a cell can the instructions contained in the nucleic acid of the virus be put into execution: the proteins of the virus are synthesized and its nucleic acid copied; the separate parts thus fashioned are then arranged as new virus particles. Once released, these particles infect other cells. A virus, therefore, does not multiply by growth and division like a cell, but by the independent production of its constituents which are finally assembled to re-constitute a new virus particle. Clearly a virus has some of the properties of a living system, but not all. It can spread, multiply and undergo mutations. It can direct the production of viral proteins that influence the environment in its favour. It is therefore subject to evolution by natural selection. But conversely, it can carry out its genetic programme and reproduce only in an environment that can already perform metabolic operations, produce energy and synthesize polymers; in other words, in a cell. A virus cannot, therefore, be considered as an organism. Outside the cell, the virus particle is but an inert object. Only the cell–virus system has all the properties of life. Viral infection is the rupture of cellular order as a result of invasion by a foreign chemical message.

There are many kinds of viruses of different shapes and sizes. In the smallest ones, the nucleic acid contains at most a few thousand base units, just enough to determine the sequence of three or four protein chains. The protein shell of these small viruses, shaped like rods or spheres, is built of several protein molecules of the same species. The way these identical molecules are assembled determines the architecture of the virus. The proteins of the viral shell can be isolated and purified. When placed in solution under certain conditions, the protein molecules aggregate by a process similar to crystallization and produce particles exactly the same shape as the virus. When the protein solution also contains molecules of the specific nucleic acid, the virus is reconstituted and the particles formed are infective. Again, the special structure of the protein, the particular way its atoms are arranged, makes it able to associate selectively with other identical protein molecules, in a structure with strictly defined symmetry and shape. It is through certain chemical patterns that the molecules of this protein recognize each other and assume a unique geometrical order. No mould is needed to establish this shape, no source of energy, no particular force, except those interactions between groups of atoms by which crystals in the inanimate world are organized and grow. The same interactions operate to produce the shape of more complex viruses formed, for instance, by a polyhedral head joined to a long tail. Several species of proteins, and not just one, must this time be arranged and fitted together in repeated symmetrical patterns to build up the complex architecture containing the nucleic acid. The same applies to organelles of the cell, such as the cytoplasmic particles where the nucleic acid sequences are translated into protein sequences and the proteins are assembled. It also applies to the flagella which decorate the periphery of the small bacterial cell.

Many of the forms we admire in cells exhibit the same properties as crystals. Crystallization implies a union of like units, a geometry strictly ordered by the forces which arrange and unite identical molecules. Whether particles, layers, fibres or tubules, most structures seen under the microscope show these characteristics. We do not yet know the molecular organization of the most complex organelles in the cell, particularly the membrane. We are now certain, however, that the construction of these organelles does not require any mysterious principle, any force unknown to physics, any factor not contained in the structure of the constituents themselves. The variety and the beauty of forms, all that geometry of living beings that fills us with wonder, seem indeed to be founded on a phenomenon which has long been known: the formation of crystals. Once again, between the living and the inanimate world, there is a difference, not in nature, but in complexity. Once again, the integration of the bacterial cell is based solely on the properties of certain proteins, on their ability to recognize other molecular species selectively. The only factor that gives unity to such a complex system is the way the constituents are coordinated. Formed of several thousand molecular species, the centre of several thousand chemical reactions that take place simultaneously at high speed, the bacterial cell could not be a functional whole without a close cohesion of its constituents. All exchanges of matter and energy must be defined to the last detail in order to accomplish the goal: to produce two bacteria. The small bacterial cell cannot, therefore, be a mere collection of molecular species enclosed in a bag and subject to the statistical laws that control juxtaposed but independent elements. There must be a communication network to inform constituents that are far apart on the atomic scale, and to direct particular activities in the general interest and for the common aim. At every stage of cell chemistry, regulatory circuits come into play and coordinate reactions, adjusting them to production requirements. In return for a small expenditure of energy, the cell thus adapts its work to its wants. It produces only what it needs when it needs it. The chemical factory is fully automated.

The coordination of the chemical activities in the cell first means the starting or arrest of reaction chains, according to the conditions that prevail. It means the continuous provision to the executive agents of information about their own activities, so that they can adapt to the situation. It means the establishment of interactions between constituents of different structures but of common function. Integration is possible only in so far as the strictness of the instructions laid down in the genetic programme is balanced by the flexibility of information collected regarding the local situation, the state of the system and the nature of the medium. This interplay between what must be done and what is actually being done continually determines the activity of each constituent. The existence of these interactions frees the system of its thermodynamic constraints and allows it to combat the mechanical tendency towards disorder. Coupled with the organs of perception and directing them, there have to be organs of execution, able to 'sound out' the external world, to detect the presence of certain compounds acting as signals and to measure their concentration. This role is played by certain proteins, known as regulators, whose structure gives them special properties. These proteins are, in fact, able to associate selectively and reversibly, not with one single molecular species, but with two or more which are different in nature and structure and which do not show any chemical reactivity between themselves. Only through the mediation of the regulatory protein can special interaction be set up between such compounds: left to themselves, they could only ignore each other chemically. These proteins constitute two-headed structures, so to speak: one head gives the protein the capacity of recognizing a particular chemical species and thus of performing a certain function, catalytic or otherwise; the other, of binding a totally different compound that modifies the configuration of the protein, and so changes the properties of the first head. Depending on whether or not this compound is present, on whether or not it reaches a certain concentration in the cell, the regulatory protein oscillates between two states: activity or inactivity. It is therefore this compound which, by its variations, modulates the functioning of the protein and, consequently, of the chain of reactions to which it belongs. It acts somewhat as a chemical signal that turns the protein 'on' or 'off'. These regulatory proteins, therefore, play the part of couplings between the different functions of the cell, between the thousands of reactions that cooperate in stocking or mobilizing energy and chemical potential. These interactions are important because they are free from the main constraints affecting chemical reactions: since they are reversible and do not involve true chemical bonds, they bring into play only activation-energies that are weak or zero; since they are not subject to rules of affinity and chemical reactivity, they can bring about couplings between any molecular species. The nature of the interactions, the very fact that they can even exist, depends entirely on the organization of the protein molecule, that is ultimately on a nucleic-acid sequence. Without these coupling agents, without these multi-headed proteins, the coordination of the cell would encounter insurmountable difficulties from the point of view both of chemical structures and of thermodynamics. The mere presence of these structures establishes a communication network between cell and environment, between genes and cytoplasm and between pairs of constituents without chemical affinity.

The regulatory circuits are based on these proteins. As in electronics, the same elements can be combined in different circuits and perform the function required in different circumstances. Some regulatory proteins detect the presence of a particular metabolite in the culture medium; but the resulting effect varies according to the nature of the compound and its role in the cell economy. If it is a sugar that can provide energy, for instance, its presence immediately sets off the synthesis of the enzymes implied in its degradation. If, on the contrary, it is an essential metabolite that the cell can produce, its presence in the culture medium at once stops the formation of enzymes that take part in its own synthesis. In a chain of reactions, the executive agents are constantly kept informed, by the final product, of the result of their activity, which they adapt accordingly. It is by means of a feedback loop that each metabolite synthesized by the cell adjusts its own production.

Like relays in electronic machines, regulatory proteins react to the presence of a chemical signal only above a certain threshold. Their response comes from the oscillation of the protein between two possible states. It represents a choice between two alternatives, between activity and inactivity, on and off, yes and no. All that the protein can detect, in fact, is the presence or absence of a given compound, a specific chemical pattern. Such binary systems function at all levels of metabolism to coordinate their diversity. They continually intervene to adapt functions to the needs of the cell and to the state of the environment; to adjust catalytic activities in reaction chains; to decide which genes must be translated into proteins; to allow the reproduction of the chromosome once per generation, and once only; to coordinate cell division. The integration of the cell and the cohesion of its activities therefore depend entirely on the existence of these protein structures which are pure products of natural selection. All imaginable interactions can be established between the most different molecular species, provided the appropriate protein structure exists. In actual fact, it is possible to observe connections that are highly unexpected from the chemical point of view, but highly efficient from the logical one. When a new metabolic route is detected, its regulation is unfailingly found to make the cell function more efficiently and economically. It is exactly the same type of advantage as a production bonus. The logic of the regulatory systems is based on the ambition of a bacterium to produce two bacteria at the lowest price.

Whether one examines the functions, the morphology or the integration of the bacterial cell, therefore, the family of proteins always occupies the front of the stage. Among them, there are executive agents as well as elements of structure or sensing devices. Whatever their role, proteins are able to join selectively with other compounds by what chemists call 'non-covalent' associations which do not involve true chemical bonds. Their special position in the cell is due to their ability to recognize specific chemical patterns in a mixture of compounds, however complex. So proteins can, as it were, 'feel' the chemical species, 'sound' the composition of the medium, 'perceive' specific stimuli of all kinds. They choose their associates because they 'know' only them. At all levels, proteins function like Maxwell's demons, fighting the mechanical tendency towards disorder. They hold the 'knowledge' by which the organization of the cell is maintained.

The very existence of each interaction results from the unique structure of the corresponding protein. This means that the anatomy of the bacterial cell, like its physiology, depends almost entirely on the details of a few protein sequences. It also means that reproduction of these structures, and hence of the whole system, is possible only because the three-dimensional molecular order is entirely determined by another, one-dimensional order. The difficulty of copying architecture in space had already been pointed out by Buffon. But this difficulty is found at all levels, with visible forms or with molecules. Molecular biology has replaced the model of an internal mould with that of a linear message. For among all material structures, it is undoubtedly the sequence that can be reproduced most exactly and at least cost. Whether considering the arrangement of characters in an organism, or of atoms in a molecule, there is only one way of envisaging the reproduction of an object: the placing of each part in the copy must be guided by the corresponding part in the original. There is no possibility of reproducing something unless every peculiarity can be marked, unless every pattern and significant detail can be recognized. This is not the case with a three-dimensional structure: only the surface is accessible, not the inside. Growth is then restricted to piling up new components on the accessible parts of the structure. This very mechanism controls the formation of a crystal by monotonously repeating the same pattern. It is quite easy, on the contrary, to imagine the reproduction of a surface, in which no detail is hidden. In theory, there is nothing to prevent the exact reproduction of a two-dimensional matrix, for example; or better still, of two matrices, each the mirror image of the other. In practice, nevertheless, it is far more complex to copy this kind of structure than a mere sequence. What makes the reproduction of molecules so efficient, what perhaps even made it possible, is the existence of a univocal relation between two systems of order: one, the double nucleic-acid sequence that always remains linear and therefore re-copies itself without difficulty; the other, the protein sequence that spontaneously and unambiguously converts itself into a specific three-dimensional structure. Complexity in space can reproduce itself only because of the underlying simplicity of a sequence; because, in the living world, the order of order is linear.

Copy and Error

In principle, the number of possible nucleic-acid sequences, and therefore of protein structures, is almost unlimited. Little is yet known about the restrictions of structure that may be imposed by certain physical constraints. But in any case, what is possible far exceeds what is achieved. The small bacterial cell contains only a few thousand protein species; but each is perfectly adapted to its function; each contributes, with an accuracy and a yield that amaze us, to the organism's project of producing two organisms. The investigation of the bacterial cell reveals today the result of consecutive reproductions over two thousand million years, the cross-section of the chain of evolution made by the present epoch. Evolution, of course, operates neither on the constituents of the bacterium nor even on the bacterium as a whole, but on large populations. Nevertheless, the very nature of its mechanism leads evolution to pay attention to details, putting each molecular structure to the test, always adapting it more closely to its function. With organisms as simple as bacteria, the only criterion for natural selection is the rate of multiplication. And in the race to produce offspring, everything has its cost: the slightest lapse, the slightest delay, the slightest divergency, provided they are hereditary. It is a law without reprieve or mercy. However minor, any difference in structure, and thus in function, almost inevitably has an evolutionary repercussion when repeated at each generation. If a structure appears that can bind several metabolites, if some circuits are thus established that can contribute to regulation, then reproduction of such a modified chromosome is immediately favoured. Natural selection chooses among organisms in existence. But *a posteriori*, everything happens as if she had chosen one by one the chemical species that make up the bacterium; as if she had fashioned each molecule and put the finishing touches to each detail.

In the genetic programme, therefore, is written the result of all past reproductions, the collection of successes, since all traces of failures have disappeared. The genetic message, the programme of the present-day organism, therefore, resembles a text without an author, that a proof-reader has been correcting for more than two billion years, continually improving, refining and completing it, gradually eliminating all imperfections. What is copied and transmitted today to ensure the stability of the species is this text, ceaselessly modified by time. Time, in this case, means the number of consecutive copies of the message, the number of successive generations leading from a remote ancestor to our present-day bacterial cell. It is probable that we shall never know the detailed paths followed by evolution. It is unlikely that we shall ever be able to discover each stage by which, perhaps over several thousand millions generations, atoms gradually became organized to constitute the formidable structure of the bacterial cell. We can be certain that at least some of the mechanisms revealed by genetics and molecular biology do take part in the variation and evolution of living organisms. It is difficult, however, to define their role and assess their importance.

Perhaps what astonishes us most in the operations performed by the bacterial cell is their accuracy. Thousands of reactions take place with a precision and exactitude far beyond anything technology and industry can accomplish. There even exist special mechanisms in the cell for detecting and correcting poor workmanship. The maintenance of the system, its constancy through succeeding generations, depends on this precision, which preserves the organism from the disintegration that inevitably menaces any mechanical system. On the whole, failures sufficiently important to cause the death of the cell, or rather its inability to reproduce, do not affect one bacterium in a thousand during growth. But exactitude does not mean infallibility. Here and there a few errors manage to creep in. These fall into two groups, according to whether they are, or are not, transmitted to descendants, whether they alter the genetic instructions themselves or interfere with their execution. In the latter case, errors may slip into the operations of transcription and especially of translation, which involve complex equipment. For instance, a protein sub-unit may be placed accidentally opposite a nucleic-acid triplet with a different meaning in the dictionary. Although the error lies in a single symbol, the sequence may produce a modified protein unable to perform the required function. Errors of this kind are rare, however, since a protein can be purified and the exact sequence of its units established without any ambiguity being detected. Such errors are not in general dangerous to the cell, for it does not matter if it produces one or two defective enzyme molecules out of the thousand or so which exercise the same function. These are merely faults of production, accidents of no consequence to the species.

Only those errors which cause a change in the genetic message itself – that is, mutations – can have important consequences for the species, for once they have appeared, they are faithfully copied in turn from generation to generation. Mutations, their origin and the way they are expressed can be investigated in bacteria more easily and efficiently than in other organisms. It is possible, in fact, to exert drastic selective pressure on enormous populations, to modify them in a particular direction. The geneticist can thus make the rarest mutants emerge. He merely has to put the several billion organisms contained in a drop of culture medium under conditions in which only the desired mutant can multiply. It is quite easy, then, to measure the frequency of mutations, to define their mechanism, to seek a possible relation of cause and effect between function and structure. The nature of mutations is imposed by the very organization of the chemical text. There is mutation when the meaning of the text is altered, when a modification occurs in the nucleic-acid sequence prescribing a protein sequence, and therefore a structure fulfilling a function. Mutations result from errors similar to those which a copyist or a printer inserts into a text. Like a text, a nucleic-acid message can be modified by the change of one sign into another, by the deletion or addition of one or more signs, by the transposition of signs from one sentence to another, by the inversion of a group of signs – in short, by anything that disturbs the pre-established order.

What characterizes these events is that they cannot be oriented in any particular direction, either by environment or by any constituent of the cell. Changes in the chemical text appear, not by modification of a previously chosen sequence, but blindly. By using certain reagents or radiations known to transform a particular chemical residue, it is possible to affect selectively one of the four nucleic-acid sub-units, to provoke the change of one letter of the text to another, a B into an A, for instance. But B is represented by millions along the chain; it is present a great many times in every segment of the message, in every gene. The chemical reaction cannot choose one particular B; it acts at random and any one of the millions of copies of B is changed into A. The only molecule that has to know the nucleic-acid sequence so as to establish its order during copying is nucleic acid itself. In the processes of protein synthesis, information is always transferred unidirectionally from nucleic acid to protein, never in the opposite direction. There exists no molecular species in nature that can modify the nucleic-acid sequence in a deliberate manner, whether among the enzymes used for ordering the nucleic acid sub-units in the copy process or among the regulatory proteins that turn nucleic-acid segments 'on' or 'off'. These molecules can establish associations with nucleic acid; but they cannot modify the sequences of the message. By the very nature of the genetic material and its relations with other cell constituents, no molecular species is able to change the plan that decides its own structure. This means that a gene cannot be transformed by reference to the function it controls. Whether spontaneous or artificially induced, mutations always modify at random the order of a sequence taken at random from the whole genetic programme. The entire system is arranged so as to make mutations blind errors. There is no component in the cell for interpreting the programme as a whole, for even 'understanding' a sequence and modifying it accordingly. Those components which translate the genetic text only understand the meaning of triplets taken separately. Those components which might change the programme when they reproduce it do not understand it. If there were a will to modify the text, it would have no direct means of action. It would have to go a roundabout way via natural selection.

Each mutation, each error in copying, affects one or more symbols in the genetic text. Each modifies one or more genes. Hence a change in one or more proteins. Depending on how 'useful' these structures are, on whether or not they favour the bacterium's purpose of reproducing itself, the mutant will or will not have an advantage compared with its congeners. Obviously, all structures and functions found today in an organism placed under conditions similar to its natural environment have been refined by selection during millions of generations. They are so precise and efficient that changes in programme can generally cause only a diminution, or even a loss, of function. Only when the environment is drastically transformed can mutation be of some benefit to the organism. Even if some thousand millions of individuals perish because they have become unable to multiply, the reproduction of a few mutants is enough for the species to adapt to new conditions. Harmful under ordinary conditions, certain mutations become advantageous in exceptional situations.

Most mutations involve changes in quality, not in quantity. They disturb the order of the genetic text, but add nothing to it. Evolution, however, occurs by an increase in the complexity of the organism, and thus by an enlargement of the programme. Two kinds of events can increase the content of genetic information in the bacterial cell. First, it sometimes happens that, when a chromosome is being reproduced, the same segment is copied twice. This is like an error in type-setting, when a line is inadvertently repeated. From then on, the two samples of the genetic fragment are perpetuated from generation to generation. The same protein is produced by the two copies of the same gene. But the selective constraints for maintaining function affect only one of the gene copies. The other copy can vary at leisure, and can mutate freely. These variations need no longer be advantageous to the cell in order to be perpetuated. It is enough that they are not harmful. Most probably, it is through such repetitions that, little by little, link by link, the chains of chemical reactions were gradually formed.

Some bacteria also have another way of adding to their genetic programme. Most frequently, micro-organisms are isolated from each other. They do not communicate. They exchange nothing. They are even protected from any relationship by their cell wall. Nevertheless, transfers of genetic material from one cell to another sometimes occur, either by the intermediary of a virus, or by processes recalling the sexuality of higher organisms. But this addition of genetic material has a lasting effect on the descendants of the cell only in so far as the fragments thus introduced succeed in taking root there, reproducing and being transmitted from one generation to the next. Such an implantation often occurs by genetic recombination. So a segment of chromosome can be replaced by an homologous segment from another individual. Among populations of bacteria that multiply under different conditions, different genetic sets tend to be formed according to environmental requirements. Through recombination, the elements of genetic texts, genes from different individuals, can be reassorted in new combinations that sometimes offer advantages for reproduction. Even though sexuality is not really a method of reproduction for bacteria, that usually multiply by fission, it nevertheless allows the different genetic programmes of the species to be mixed with the resultant appearance of new genetic types.

Recombination only reassorts the genetic programmes in populations; it does not add to them. Certain genetic elements are, however, transmitted from cell to cell and simply added to the genetic material already present. The instructions they contain are indispensable neither for growth nor for reproduction. But this addition to the genetic text allows the cell to acquire new structures and perform new functions. It is an element of this type that determines sexual differentiation in certain species of bacteria, for instance. Furthermore, as it is not indispensable, the nucleic-acid sequence contained in such supernumerary elements is not subject to the constraints of stability that natural selection exercises on the bacterial chromosome. These elements represent a free addition for the cell, a sort of reserve of nucleic-acid text that can vary freely in the course of generation.

Two apparently opposed properties of living beings, stability and variability, are based on the very nature of the genetic text. At the level of the individual – the bacterial cell – one observes the recopying, with extreme rigour, of a programme which prescribes not only the detailed plan of each molecular structure, but the means of executing the plan and of coordinating the activities of the structures. On the other hand, at the level of the bacterial population, or of the species as a whole, the nucleic-acid text appears to be perpetually disorganized by copying errors, by recombinational spoonerisms, by additions or omissions. In the end, the text is always rectified. But it is rectified neither by a mysterious will seeking to impose its design, nor by an environmentally determined reordering of the sequence: the nucleic-acid message does not learn from experience. The message is rectified automatically by a process of selection exerted, not on the genetic text itself, but on whole organisms, or rather populations of organisms, to eliminate any irregularity. The very concept of selection is inherent in the nature of living organisms, in the fact that they exist only to the extent they reproduce. Each new individual which by mutation, recombination and addition becomes the carrier of a new programme is immediately put to the test of reproduction. If this organism is unable to reproduce, it disappears.

If it is able to reproduce better and faster than its congeners, this advantage, however minor, immediately favours its multiplication and hence the propagation of this particular programme. If in the long run the nucleic-acid text seems to be moulded by environment, if the lessons of past experience are eventually written into it, this occurs in a roundabout way through success in reproduction. But only what exists is reproduced. Selection operates, not on possible, but on existing, living organisms.

Unlike the structure of the genetic text, the execution of the prescriptions it contains is subject to specific influences of environment. There again, however, environment does not give instructions. In certain processes, such as induced synthesis of enzymes, the bacterial cell responds to the presence of specific compounds in the environment by producing specific proteins. Only a few years ago, it appeared inevitable that the compound had in some way to adjust the meaning of the genetic sentence, to give it its share of the order, to contribute to the final structure of the protein. This is now known to be wrong: even in these phenomena, environment does not give instructions. The specific compound plays the part of a simple stimulus; it merely initiates a synthesis in which the mechanisms and structure of the end product are strictly fixed by the nucleic-acid text. The system only offers a choice between two alternatives. The only instruction that can be received from the environment through regulatory proteins is a 'go' or 'stop' signal. Reading the genetic message, therefore, is like getting music from a juke-box in a café. By pressing one of the buttons, one can choose the desired record from those in the machine. But in no case can one modify the recorded music or its execution. Likewise a segment of the genetic text contained in the bacterial chromosome may or may not be transcribed, depending on the chemical signals received from the environment; but the signals cannot modify its sequence, and therefore, its function. The old word 'adaptation' thus covers two different things. On the one hand, it means a phenomenon that occurs in an individual; in some way it expresses the response of the organism to some external factor, but always within the limits set by the instructions contained in the programme. On the other hand, it means the modifications that occur in a population; this involves a change in the programme itself under the effect of a selective pressure that favours certain programmes as they appear. But whether it is a question of exploiting the possibilities of an existing programme or of changing it, adaptation is always the result of a selective, not an instructive, effect of environment.

*

This is how molecular biology sees the threshold of integration separating the living world from the inanimate world. Of course, we are still far from knowing all the details of an object as simple as a bacterium. Of the two thousand or so reactions that take place in the tiny chemical factory, only six or seven hundred have been identified and studied to date. We do not know the composition and structure of many chemical species that recognize each other and assemble in organelles. The composition and structure of the membrane, for instance, are still hardly known. But while each step forward shows how complex the details are, it also shows how simple are the principles involved. The procedures used by nature strongly resemble those employed in human technology, whether for lengthening polymers, transferring information or gearing regulatory circuits. As the bacterial factory is studied, its mechanisms taken to pieces and its structures investigated, nothing emerges that, in theory at least, does not appear within the scope of experimental chemistry: no gap between the behaviour of small inorganic molecules and that of enormous organic structures, between the reactions of the inanimate and of the living, between chemistry in the laboratory and the chemistry of organisms. Enzymic activity is revealed in solution in a test-tube and affects molecules that can often be produced in the laboratory; all the interactions involved are known to physicists. Recently an enzyme has even been synthesized that has the same catalytic activity as its natural counterpart. In theory, such an operation presents no major difficulty; chemists have long known how to make in the laboratory the bond that joins two amino-acid residues in a protein chain. In practice, however, this synthesis is an extremely complex operation: several hundred sub-units have to be aligned, then linked end to end. This is done by adding one sub-unit at a time, each operation requiring a series of chemical reactions. The protein is formed at a cost of several thousand reactions that have to be carried out one after another in fixed order. Although machines have been constructed that can automatically perform some parts of the work in the order prescribed by a programme, this kind of synthesis is still difficult, time-consuming and of low yield. But one can safely predict that a variety of proteins will soon be produced; not simply molecules with either a catalytic or some other activity, but also unknown structures with no function; something like the kind of protein that must have been formed on earth before living beings appeared. These structures with no role or purpose would make it possible to explore a past hardly accessible to experimentation.

Not on principle, but for technical reasons, it is even more difficult to produce a long nucleic-acid chain with an exact sequence in the laboratory. Not because chemists do not know how to hook the nucleic sub-units together one by one; but because the known procedures are still quite inefficient. Even if each reaction proceeded with a yield of more than ninety per cent – and we still are far from that figure – after the thousands of operations necessary, the quantity of product obtained would be absurdly small compared with the raw materials used. Yet short sequences of a few sub-units can be synthesized and then repeated in long chains. The synthesis of a gene has already been achieved by methods that combine organic chemical procedures and the use of bacterial enzymes. In all likelihood, it will become possible to make nucleic-acid sequences which can take root in a cell and confer on it some new function. What appears to be out of reach, and for a long time yet to come, is the creation, piece by piece, of a long nucleic-acid chain, the elaboration of a genetic programme by synthesis, even as simple a programme as that of a virus. But there is no *a priori* reason why the chemist should not eventually succeed in this enterprise.

Many physicists were attracted to biology, and more especially to the investigation of heredity, by the hope of finding some new law that had not hitherto been revealed in the study of matter. Their expectations were not fulfilled. Certainly, for technical reasons, complex structures will remain beyond the reach of experimental chemistry for a long time to come. It is not inconceivable, however, that in the future the thousands of chemical species contained in the bacterial cell may be synthesized one by one. But there is no chance of seeing *all* these compounds being assembled correctly and a bacterium emerging fully armed from a test-tube. Many enzymes that polymerize units function only in the presence of a matrix. Others can only lengthen fragments of pre-existing polymers that are used as primers. During the formation of certain complex organelles, the correct orientation of the sub-units might even be sometimes determined by the presence of similar structures already formed, as in the formation of crystals. However rudimentary the bacterial cell may appear in comparison with other living organisms, it still required a considerable period of time for its system to be organized. A bacterium functions so effectively because for two million years or more its ancestors were trying their hand at this chemistry, scrupulously recording the recipe for each success. This is where the division comes between the living world and the inanimate world, between biology and physics. Inanimate bodies do not depend on time. Living bodies are indissolubly bound up with time. In the living world, no structure can be detached from its history.

The history of our bacterium is really a continuous chain of reproductions, with their adventures, their failures, their successes. There are bacteria on earth today because in the course of time other bacteria, or something even more rudimentary, tried desperately to reproduce times out of number. As the copying mechanism is not infallible, there were many opportunities to change, damage and improve the system. Evolution is built on accidents, on chance events, on errors. The very thing that would lead an inert system to destruction becomes a source of novelty and complexity in a living system. An accident can be transformed into an innovation, an error into success. For natural selection is a game with its own rules. All that count are the changes that affect the number of offspring. If they reduce that number, they are mistakes; if they increase it, they are exploits. There are neither tricks not stratagems in the game; only a careful score of profit and loss. Reproduction directs the course of chance.

The little bacterial cell is so arranged that the whole system can reproduce as often as once every twenty minutes. With bacteria, unlike organisms which reproduce only sexually, birth is not counterbalanced by death. When bacterial cultures grow, the individual bacteria do not die. They disappear as individuals: where there was only one, suddenly there are two. The molecules of the 'mother' are distributed equally among her 'daughters'. For instance, the mother contained a long duplex of deoxyribonucleic acid that splits into two before cell division. Each daughter receives one of these identical duplexes, each of which is formed by an 'old' chain and a 'new' one. One of the criteria that a bacterium is no longer alive is its inability to reproduce. If this non-life is to be seen as death, it is a contingent death. It often depends on the conditions of the culture. When a small part of a culture is continuously replaced by fresh medium, such a culture remains in a state of perpetual growth: bacteria go on reproducing indefinitely.

What makes an individual ephemeral in a bacterial population is not, therefore, death in the usual meaning of the word, but dilution entailed by growth and multiplication. Only the organization persists, and that is automatically reproduced as long as the cell can obtain energy and materials from the medium. There is no Mind to direct operations, no Will to order them to continue or to stop. There is only the perpetual execution of a programme that cannot be dissociated from its fulfilment. The only elements that can interpret the genetic message are the products of the message itself. The genetic text makes sense only for the structures it has itself determined. There is thus no longer a cause for reproduction, simply a cycle of events in which the role of each constituent is dependent on the others. Organization could reproduce itself and living organisms emerge, because the complexity of structures in space happened to be generated by the simplicity of a linear combinative system; and also because a univocal relation was established between two systems of symbols: one maintaining information through successive generations, and the other unfolding the structures at each generation. The first system sets up vertical communication from parent to child; the second determines horizontal communication between constituents of the organism. The relation between these two systems gives cell reproduction an internal logic that no intelligence has chosen. But this very logic excludes the genetic message from any change directed either by the environment or by the cell itself. Only the activity of the genetic material, not its structure, is subject to the regulation that coordinates the components of the organism. This does not in any way mean that the nucleic message is free of all supervision. Although it is not subject to control inside the cell, the genetic programme is still subject to regulation from outside; not because a mysterious hand guides the destiny of each bacterium, but because an individual is part of a population, whether living in a test-tube, a puddle of water or a mammal's intestine. It constitutes, therefore, a simple component in a system at a higher level that obeys another logic. Inside this system is a regulation that affects the genetic programme. The interactions between the population and its environment eventually have repercussions on the reproduction of the novelties that occur spontaneously in the genetic text. In the last analysis, there does exist, between the genetic programme and the environment, that necessary relation required by adaptation. But this relationship is only established in a roundabout way through a long feedback loop that adjusts the quality of the message according to the number of descendants. The genetic text is constantly re-arranged, and the message is ceaselessly modified, corrected and adapted to reproduction under the most varied conditions, because of the successive revisions carried out by natural selection. Without any thought to dictate it, without any imagination to renew it, the genetic programme is transformed as it is carried out.

Conclusion

The Integron

That heredity can today be interpreted in molecular terms does not constitute an end in itself; nor is it a proof that all biology must in future become molecular. It signifies primarily that the two major currents of biology, natural history and physiology, that went their separate ways for so long almost unaware of each other, have finally joined forces. The old quarrel between integrationists and reductionists has thus been resolved in the distinction established long ago by physics between the microscopic and the macroscopic. On the one hand, the variety of the living world, the wonderful diversity of forms, structures and properties at the macroscopic level are based on the combinative system of a few molecular species, that is, on very simple devices at the microscopic level. On the other hand, the processes that take place at the microscopic level in the molecules of living beings are completely indistinguishable from those investigated in inert systems by physics and chemistry. Only at the macroscopic level of organisms do special properties appear, imposed by the necessity of self-reproduction and of adaptation to certain conditions. The problem, then, is to interpret the processes common to beings and things in terms of the special status assigned to living organisms by their origin and purpose.

Recognition of the unity of physical and chemical processes at the molecular level has deprived vitalism of its *raison d'être*. In fact, since the appearance of thermodynamics, the operational value of the concept of life has continually dwindled and its power of abstraction declined. Biologists no longer study life today. They no longer attempt to define it. Instead, they investigate the structure of living systems, their functions, their history. Yet at the same time, recognition of the purpose of living systems means that biology can no longer be studied without constant reference to the 'plan' of organisms, to the 'sense' which their very existence gives to structures and functions, an attitude obviously very different from the reductionism that was long dominant. In the era of reductionism, to be really scientific, analysis had to exclude any considerations beyond the system immediately under study and its specific role. The rigour imposed on description required elimination of that element of finality which the biologist refused to admit. Today, in contrast, one can no longer separate a structure from its significance, not only in the organism, but in all the chain of events that have led the organism to become what it is. Every living system is the result of a certain equilibrium between the parts of an organization. The interdependence of these parts means that modification at any point affects the whole of relationships and sooner or later produces a new organization. By isolating systems of different kinds and complexity, it is possible to recognize their constituents and justify their relationships. Yet whatever the level studied – molecules, cells, organisms or populations – the perspective is necessarily historical and the principle of explanation necessarily that of succession. Each living system has to be analysed on two planes, two cross-sections, one horizontal and the other vertical, which can be separated only for the sake of explanatory convenience. On the one hand, one has to distinguish the principles governing the integration of organisms, their construction, their functioning; and on the other, the principles that directed their transformations and their succession. The description of a living system requires reference to the logic of its organization, as well as to the logic of its evolution. Today biology is concerned with the algorithms of the living world.

*

The organization of living systems obeys a series of principles, as much physical as biological: natural selection, minimum energy, self-regulation, construction in 'stages' through successive integration of sub-sets. Natural selection imposes finality, not only on the whole organism, but on each of its components. In a living organism, every structure has been selected because it fulfils a certain function in a dynamic self-reproducing whole. It is therefore by their history and continuity that the molecules composing living systems differ from all others. Some have not varied for millions of years: in a certain sense, they remain copies of the molecules formed in ancient times. Others, on the contrary, have been transformed under some selective pressure. Numerous were those lost on the way; perhaps more numerous still those which appeared in new species, in man, for instance. But over and above the demands of selection, living systems, just like inanimate systems, remain subject to the law of minimum energy. Whether or not they involve true chemical bonds, whether they entail syntheses or mere associations of molecules, the reactions in the living organisms always proceed in the same direction, towards a decrease of free energy. The rate of these reactions is always determined by the activation energies required by the transitions involved.

Regulatory circuits give living systems both their unity and the means of conforming to the laws of thermodynamics. These laws state that a chemical reaction can be modified only in terms of its equilibrium or rate. In a simple reaction, the equilibrium constant is a function of the molecules taking part. Catalysis simply increases the rate by decreasing the required activation energy. In an enzymic reaction involving such a complex structure as a protein, it is the shape of the protein that determines both its affinity for the substrate and the rate of the reaction. Affinity and rate can be changed only by changing the shape of the protein. All the coordination of the cell is thus dependent on the geometrical deformation of a few proteins by interactions with certain metabolites acting as specific signals. In multicellular organisms, there are additional regulatory circuits for harmonizing and integrating the activities of the cells. Here direct contacts between cells, as well as hormones and the nervous system, play their part. It is not yet known how these circuits function. It seems likely, however, that hormones and chemical mediators of the nervous system also act by deforming certain proteins in the membrane or the cytoplasm of sensitive cells. In themselves, these compounds have no significance. They acquire value as signals for certain cells only because of the presence of proteins that serve as receptors, i.e. ultimately because of the genetic programme in these cells. But in all cases the regulation of biological systems affects the equilibrium and rate of reactions. In all cases, it simply expresses the interaction of components, that is, the properties inherent in their arrangement, and therefore in their structure.

Construction in successive stages is the principle governing the formation of all living systems, whatever their degree of organization. Even the simplest organism is so complex that it could probably never have taken shape, reproduced and evolved if the whole had had to be built piece by piece, molecule by molecule, like a mosaic. Instead, organisms are built by a series of integrations. Similar elements are assembled to form a set of the level just above, and so on. It is thus by combining more and more elaborate elements, by fitting subordinate structures into one another, that complexity is born in living systems. These systems can be reproduced from their elements at each generation, because at each level the intermediate structures are thermodynamically stable. Living beings thus construct themselves in series of successive 'parcels'. They are arranged according to a hierarchy of discontinuous units. At each level, units of relatively well defined size and almost identical structure associate to form a unit of the level above. Each of these units formed by the integration of sub-units may be given the general name 'integron'. An integron is formed by assembling integrons of the level below it; it takes part in the construction of the integron of the level above.

This hierarchy of integrons, this principle of a box made up of boxes is already illustrated at the microscopic level in the way protein structures are produced inside the cell. Three stages, in fact, can be seen in the building of these structures. In the first stage, inorganic elements are converted into small specific molecules, the protein sub-units, or amino acids, by a series of enzymic reactions. The specificity of the reactions depends on the associations between enzymes and substrates and on their equilibria. Their rates are coordinated through the interaction of enzymes with certain metabolites. In a second stage, the polymers are arranged along nucleic-acid templates where protein sub-units are lined up in precise order. This arrangement depends on specific associations that do not involve any chemical bond. Only when they are properly in place are the sub-units connected to each other by enzyme action. In a third and final stage the protein chains fold up and form superstructures.

The simplest superstructures are assembled only as a result of the capacity of association that their structure affords to the components: the affinity of the elements for each other is sufficient for the system to form spontaneously. For more complex superstructures, kinds of 'centres' are perhaps involved in the organization of some components; centres acting either as structural agents to modify the conformation of the other components, as kinds of enzymes to accelerate their association, or even as templates favouring one particular arrangement among all those that are thermodynamically permissible. But invariably, the possible arrangements of an organized structure depend on the bond energies between elements. Invariably, they are an equilibrium property of the system. Even if such centres exist, their formation is still determined by the interactions of the components. In the end, the most complicated structures are built up in a series of stages in which intermediates may be used, not only as material, but also, should the occasion arise, as agents in constructing the next structure. Until further notice, only the components incorporated into the structure are required for its formation. Living organisms are formed by spontaneous assembly of their components.

In many ways the properties of these structures recall those of crystals. This is an old analogy, already invoked more than two centuries ago to explain the shape, growth and reproduction of organized beings. It had been necessary to abandon this comparison, however, once the structure of a perfect crystalline solid was brought to light. Such a crystal requires the same pattern to be repeated in three dimensions. It is a regular arrangement of atoms from the centre to the surface. Being inaccessible, the interior of the structure has no function. The crystal can develop only by the addition of components to its surface. It does not reproduce. But subsequently, the concept of a crystal has been generalized: it applies to any organization of matter that is repetitive in two, or even in one dimension. From particles that have no dimension, so to speak, fibres and surfaces like membranes or three-dimensional bodies can form spontaneously. From this point on, the analogy between crystals and living structures regains an operational value. What gives a collection of objects the property of assembling is their sameness.

Not only can they form geometrical structures; they can do so spontaneously. But there is no way of telling how far the sameness must go and what differences in structure can be tolerated. Although constraints on the formation of three-dimensional crystals appear strict, they seem less stringent in other cases, so that nucleic-acid or protein sub-units are sufficiently similar objects to be placed in geometrical arrangements. A whole series of biological structures – polymers, membranes and intracellular organelles – thus have their own internal logic, a logic which is not exactly that of the three-dimensional crystals, but very little different. All these structures can exercise a chemical function only through their surface.

Yet, although the principles involved in the organization, construction and logic of living systems can now be perceived, although their origin can be glimpsed by extrapolation, it is still hard to grasp the series of events that led from the organic to the living. For the biologist, the living begins only with what was able to constitute a genetic programme. For him, an object deserves the name of organism only when it offers a foothold for natural selection. He sees the mark of the living in the ability to reproduce, even if a primitive organism may have required several years to form its like. For the chemist, in contrast, it is somewhat arbitrary to make a demarcation where there can only be continuity. Every organism contains a panoply of structures, functions, enzymes, membranes, metabolic cycles, energy-rich compounds and so on. Whatever the beginning assigned to what is called a living system, it is possible to envisage its organization only in an environment already prepared well in advance. Biological evolution is necessarily the unbroken continuation of a long process of chemical evolution. It is possible to try to reconstitute in the laboratory the conditions that apparently prevailed on earth before the appearance of living organisms. Whole series of organic compounds are then seen to form spontaneously. Even polymers can arise by chance associations between the sub-units. Although inefficient, the reactions required for producing the macromolecules characteristic of living organisms really seem to occur without biological catalysts. Yet it is difficult to imagine the appearance of an integrated system, however primitive; the origin of an organization able to reproduce even badly, even slowly. For the humblest organism, the simplest bacterium, is already a coalition of enormous numbers of molecules. It is out of the question for all the pieces to have been formed independently in the primeval ocean, to meet by chance one fine day, and suddenly arrange themselves in such a complex system. The first ancestor could only have been some kind of nucleus, an association of several molecules helping each other to re-form after a fashion. But then how did it all begin? And with what? The genetic message can be translated only by the products of its own proper translation. Without nucleic acids, proteins have no future. Without proteins, nucleic acids remain inert. Which is the hen, which the egg? And where can traces be found of this precursor, or of some precursor of the precursor? In some still unexplored corner of the globe? On a meteorite? On another planet of the solar system? Without any doubt, the discovery somewhere or other, if not of a new form of life, at least of somewhat complex organic vestiges, would be priceless. It would transform our way of envisaging the origin of genetic programmes. But as time passes, the hope of this diminishes.

For want of vestiges to examine, biology is reduced to making conjectures. It tries to arrange the problems in series, to individualize the objects and formulate questions that can be answered by experiments. Which of the polymers, nucleic acid or protein, came first? What is the origin of the genetic code? The first question leads one to speculate whether anything vaguely like a living organism would be conceivable without both types of polymer. The second raises problems both of evolution and of logic. Of evolution, because univocal correspondence between each group of three nucleic-acid subunits and each protein sub-unit cannot have arisen at a single stroke. Of logic, because it is difficult to perceive why this particular correspondence was adopted rather than another; why one nucleic-acid triplet 'means' a certain protein sub-unit and not another. Perhaps primitive organizations had some constraints of structure we know nothing about: it would then be the adjustment of molecular conformations that would have imposed, if not the whole system, at least some of its equivalences. But again perhaps there was no constraint at all: then it would have been purely by chance that the equivalences were produced and persisted afterwards. For once a system of relations has been established, the relations cannot be changed without the risk of the whole meaning of the system being lost and all its value as a message destroyed. A genetic code is like a language: even if they are only due to chance, once the relations between 'sign' and 'meaning' are established, they cannot be changed. These, then, are the questions molecular biology is trying to answer. But nothing indicates that the transition between the organic and the living can ever be really investigated. It may perhaps never become possible to estimate what the probability was of a living system appearing on earth. If the genetic code is universal, it is probably because every organism that has succeeded in living up till now is descended from one single ancestor. But, it is impossible to measure the probability of an event that occurred only once. It is to be feared that the subject may become bogged down in a slough of theories that can never be verified. The origin of life might well become a new centre of abstract quarrels, with schools and theories concerned, not with scientific predictions, but with metaphysics.

And yet biology has demonstrated that there is no metaphysical entity hidden behind the word 'life'. The power of assembling, of producing increasingly complex structures, even of reproducing, belongs to the elements that constitute matter. From particles to man, there is a whole series of integration, of levels, of discontinuities. But there is no breach either in the composition of the objects or in the reactions that take place in them; no change in 'essence'. So much so, that investigation of molecules and cellular organelles has now become the concern of physicists. Details of structure are now defined by crystallography, ultracentrifugation, nuclear magnetic resonance, fluorescence and other physical techniques. This does not at all mean that biology has become an annex of physics, that it represents, as it were, a junior branch concerned with complex systems. At each level of organization, novelties appear in both properties and logic. To reproduce is not within the power of any single molecule by itself. This faculty appears only with the simplest integron deserving to be called a living organism, that is, the cell.

But thereafter the rules of the game change. At the higher-level integron, the cell population, natural selection imposes new constraints and offers new possibilities. In this way, and without ceasing to obey the principles that govern inanimate systems, living systems become subject to phenomena that have no meaning at the lower level. Biology can neither be reduced to physics, nor do without it.

Every object that biology studies is a system of systems. Being part of a higher-order system itself, it sometimes obeys rules that cannot be deduced simply by analysing it. This means that each level of organization must be considered with reference to the adjacent levels. It is impossible to understand how a television set works without first knowing how transistors work and then something about the relations between transmitters and receivers. At every level of integration, some new characteristics come to light. As physicists already observed at the beginning of the twentieth century, discontinuity not only requires different means of observation; it also modifies the nature of phenomena and even their underlying laws. Very often, concepts and techniques that apply at one level do not function either above or below it. The various levels of biological organization are united by the logic proper to reproduction. They are distinguished by the means of communication, the regulatory circuits and the internal logic proper to each system.

*

Everyone agrees that there is a direction in evolution. In spite of errors, of dead ends, of false starts, a certain road has been covered during more than two thousand million years. Yet it is difficult to describe the course that natural selection has imposed on chance. The words 'progress', 'progression' and 'improvement' are not suitable. They suggest too much regularity, purpose and anthropomorphism. Their criteria remain ill defined. If one criterion is adaptation to survive, then the bacterium *Escherichia coli* appears just as well adapted to its environment as man to his. The words complication or complexity are hardly better. There are gratuitous complications, and others that, because of over-specialization, prohibit any possibility of further evolution. What is perhaps most characteristic of evolution is the tendency to flexibility in the execution of the genetic programme; it is an 'openness' that allows the organism constantly to extend its relations with its environment and thus to extend its range of action. In so simple an organism as a bacterium, the programme is carried out with great rigidity. It is 'closed' in the sense that the organism can only receive very limited information from its environment and can only react in a strictly determined way to this information. All that a bacterium perceives is the presence or absence of certain compounds in the culture medium. The sole response that it makes is to produce or not to produce the corresponding proteins. Its perceptions and reactions are reduced to one alternative, yes or no. 'Success' in evolution leads to increases in both the ability to perceive and the ability to react. For an organism to differentiate, for it to become more independent and to extend its exchanges with the outside world, there must be a development not only of the structures which link the organism to its environment, but also of the interactions which coordinate its constituents. At the macroscopic level, therefore, evolution depends on setting up new systems of communication, just as much within the organism as between the organism and its surroundings. At the microscopic level, this is expressed by changes in genetic programme, both qualitative and quantitative.

The notion that evolution results exclusively from a succession of micro-events, from mutations, each occurring at random, is denied both by time and by arithmetic. For the wheel of chance to come up step by step, sub-unit after sub-unit, with each of the several ten thousand protein chains needed to compose the body of a mammal would require far more time than the span generally attributed to the solar system. Only in very simple organisms can variation occur entirely in small independent stages. Only in bacteria can speed of growth and size of populations allow the organisms to await for the appearance of a mutation in order to adapt. Evolution has become possible, only because genetic systems have themselves evolved. As organisms become more complicated, their reproduction also becomes more complicated. A whole series of mechanisms appears, always based on chance, which help to reassort the programmes and compel them to change: the scattering of the genetic programme over several chromosomes; the presence of not one but two copies of each chromosome in each cell; the alternating phases of one set or two sets of chromosomes during the life cycle; the independent segregation of chromosomes; the recombination by breakage and reunion of homologous chromosomes, and so on. But the most important inventions are sex and death.

Sexuality seems to have arisen early in evolution. At first it was a kind of auxiliary of reproduction, a superfluous gadget, so to speak: nothing obliges a bacterium to make use of sexuality in order to multiply. It is the necessity of resorting to sex as a reproductive device that radically transforms the genetic system and the possibilities of variations. As soon as sexuality becomes obligatory, each genetic programme is no longer formed by exactly copying a single programme, but by reassorting two different programmes. The genetic programme is then no longer the exclusive property of one line of descent. It belongs to the collectivity, the group of individuals who communicate with each other by means of sex. Thus a kind of common genetic fund is set up, drawn on by each generation for making new programmes. This common fund, this population united by sexuality, forms the unity of evolution. Instead of the sameness imposed by the strict reproduction of the programme, sexuality offers the diversity produced by a reassortment of programmes at each generation. So great is this diversity that, with the sole exception of identical twins, no individual is exactly like his brother. Sexuality obliges programmes to cover all the possibilities of the genetic combinative system. It therefore compels change. In order to convince oneself that sex plays such a role in evolution, that it is itself an object of evolution open to continuous refinement, it is enough to consider the subtleties, the rites and the complications which accompany sexual practices in higher organisms.

The other necessary condition for the very possibility of evolution is death. Not death from without, as the result of some accident; but death imposed from within, as a necessity prescribed from the egg onward by the genetic programme itself. For evolution is the result of a struggle between what was and what is to be, between the

conservative and the revolutionary, between the sameness of reproduction and the newness of variation. In organisms reproducing by
fission, the dilution of an individual caused by the rapidity of growth
is sufficient to erase the past. But in multicellular organisms, with
differentiation into somatic and germ lines, with sexual reproduction,
individuals have to disappear. This is the resultant of two opposite
forces: an equilibrium between sexual effectiveness on one hand,
with its cortège of gestation, care and training; and the disappearance
of the generation that has completed its role in reproduction on the
other. The adjustment of these two parameters by the effect of
natural selection determines the maximum duration of life of a
species. The whole system of evolution, at least in animals, is based
on such an equilibrium. The limits of life cannot be left to chance.
They are prescribed by the programme which, from the moment the
ovule is fertilized, fixes the genetic destiny of the individual. The
mechanism of ageing is not yet known. The theory at present most
favoured considers senescence as the result of accumulated errors,
either in the genetic programmes contained in somatic cells or in the
way these programmes are expressed, that is, in the proteins produced
by the cells. According to this theory, the cell might cope with a
certain number of errors, but once beyond this point, it would be
doomed to die. In time, errors accumulated in an increasing number
of cells would cause the inevitable extinction of the organism. The
very way the programme is executed would, therefore, determine
the length of life. However this may be, death is an integral part of
the system selected in the animal world and its evolution. Much may
be hoped from what today is called 'biological engineering': the
cures for many scourges, cancer, heart disease, mental illness; the
replacement of various organs with grafts or artificial parts; a cure
for some failings of old age; the correction of certain genetic defects;
even the temporary interruption of active life to be resumed at will
later. But there is very little chance that it will ever be possible to
prolong life beyond a certain limit. The constraints of evolution
can hardly be reconciled with the old dream of immortality.

The arsenal of genetics favours mainly changes in quality of the
programme, not in its quantity. In fact, evolution is first expressed

by increased complexity. A bacterium is the translation of a nucleic-acid sequence about one millimetre long and containing some twenty million signs. Man is the result of another nucleic-acid sequence, about two metres long and containing several thousand million signs. The more complicated the organization, the longer the programme. Evolution became possible, through the relationship established between the structure of the organism in space and the linear sequence of the genetic message. The complexity in integration is then expressed by the simplicity of an addition. The known mechanisms of genetics, however, favour variations of the programme but hardly ever provide it with any supplement. There are certain copying errors that repeat certain segments of the message, genetic fragments that viruses can transfer or even supernumerary chromosomes. But these processes are not very effective. It is hard to see how they could be sufficient to cause some of the major stages in evolution: the change in cellular organization from the simple or 'procaryotic' form of bacteria to the complex or 'eucaryotic' form of yeasts and higher organisms; or the transition from the unicellular to the multicellular state; or the appearance of vertebrates. Each of these stages, in fact, corresponds to a rather important increase in nucleic acid. These sudden increases can have occurred only by making the most of some exceptional chance event, such as an error in reproduction providing extra chromosomes, or even some exceptional process, such as a symbiosis of organisms or the fusion of genetic programmes from distinct species. The fact that symbioses can indeed take part in evolution is now proved by the nature of 'mitochondria', these organelles responsible for producing energy in complex cells; by all biochemical criteria, they bear the stamp of bacteria. They even have their own nucleic-acid sequence independent of the chromosomes of the host cell. In all likelihood, they are vestiges of bacteria that once associated with another organism to form the ancestor of our cells. As to fusions of genetic programmes, they are known in plants, but not in animals, which are protected by a safety mechanism from the effects of the 'abominable couplings' dear to antiquity and the Middle Ages. Cells from different species, however, have recently been fused in laboratory cultures, human and mouse cells, for example. Each possessing both the human and mouse programmes, these hybrid cells multiply perfectly. What abnormal couplings between different species cannot achieve may nevertheless be accomplished in other ways. Were such encounters able, even exceptionally, to have consequences, this is enough to provide an opportunity for very profound changes. In practice, nothing proves that such accidents occur in nature; but in theory they are not impossible. There is no regularity in expansions of programme. There are sudden changes, unexpected increases, unexplained decreases, with no relation to the complexity of the organism. Very unusual events are required to fit enlargements of programme into the rhythm of evolution. This shows how illusory any hope of estimating the duration or evaluating the probabilities of evolution are today. One day perhaps, computers will calculate what the chances were of man appearing on earth.

Expansion of programme is caused by the tendency to increase interactions between the organism and its environment, a characteristic feature of evolution. There are many ways an organism can multiply exchanges with its surroundings. Already protozoa succeed in doing so. Their outfit of specialized organelles shows a surprising degree of complexity for a single cell. But there is a limit to the number and size of structures compatible with reproduction. Beyond a certain threshold, to increase the number of cells and differentiate them becomes a way of economizing. While some cells look after nutrition, others can deal with perception, locomotion or integration. Diversifying and specializing cells means freeing each from the constraints imposed by the necessity of having to accomplish *all* the reactions of the organism. It means allowing each cell to do less, but to do it better, so long as activities are coordinated. If they are to specialize, cells must therefore communicate with each other.

There are several ways cells can communicate: by direct contact or through the mediation of the nervous system and the hormones. Little is yet known of the nature of the molecular interactions that take part in such regulatory circuits. In fact, we are beginning to 'understand' the cell, but not the tissues or organs. Nothing is known of the logic of the system controlling the execution of complex programmes such as the development of a mammal. The formation of a man from an egg is a marvel of exactitude and precision. How can millions of millions of cells emerge, in specialized lineages, in perfect order in time and space, from a single cell? This baffles the imagination. During embryonic development, the instructions contained in the chromosomes of the egg are gradually translated and executed, determining when and where the thousands of molecular species that constitute the body of an adult are to be formed. The whole plan of growth, the whole series of operations to be carried out, the order and the site of syntheses and their coordination are all written down in the nucleic-acid message. And in the execution of the plan, there are few failures: the accuracy of the system may be measured by the rarity of abortions and monsters.

During development, each cell receives a complete set of chromosomes. But according to their specialization, different cells produce different types of messengers and proteins. Although it contains the whole programme, each cell translates only part of it and carries out only certain instructions. There is thus a precise sequence of chemical events during which the very expression of the genes is modified as the cells differentiate. Through the interplay of regulatory circuits segments of the message in each cell-line are activated or inhibited. Not only are these regulatory circuits more complex in multicellular organisms than in bacteria, they also fulfil different requirements. First, because in these organisms there must be systems that can differentially activate sets of genes in a permanent, instead of reversible manner. Also, because finding one gene among a million, and not one among a thousand, requires a more elaborate mechanism, such as successive sorting of sub-sets. Finally, because a bacterium and a cell in a multicellular organism operate under very different conditions. The bacterium has to maintain its functional equilibrium while adapting to different environments. The cell must also preserve a precise state of equilibrium; but in addition, it must coordinate its activities with those of its neighbours. Only in this way can the organ fulfil its functions, which in turn are subject to regulation by the organism as a whole.

In the end, it is always the logic of the organism, its individuality, its purpose which control its constituents and their systems of communication. In the network that coordinates such a complex group of chemical activities as a mammal, however, there are many opportunities for errors or false manoeuvres. Some are without importance; others have major consequences. Cellular multiplication, for instance, is subject to control by the organism. Swift at first during development of the embryo, it ceases completely when the organism reaches maturity, only resuming in response to injury. The genetic programme does not simply prescribe the plan of cellular divisions; it also sets a limit to them. This coordinating network seems to combine two kinds of circuits: one direct, mediated by actual contact between the cells; the other indirect, mediated by hormones. In each case, however, it is through specific receptors on its surface that the cell receives the signals. Were a receptor inactivated, were a signal not transmitted, then one of the circuits ensuring the social behaviour of molecules and cells would be interrupted. A cell may thus be led into a state of anarchy: deaf to signals limiting its growth, it is no longer a member of the community. It may invade neighbouring tissues and cause a tumour. With the notion of genetic programme, the old controversies about the origin of cancer have lost much of their significance. Whether the lesion starts in the nucleus or the cytoplasm, whether it be the consequence of a somatic mutation, of the presence of a virus or of a defect in a circuit, anything which prevents reception of a signal can put the cell outside the laws of the community. To understand cancer is to gain access to the logic of the system which imposes on cells the constraints of the organism.

All these complications caused by the multiplicity of cells and by their differentiation are determined by the increased exchanges between the organism and its environment. To heal a wound after injury, or to regenerate a limb after amputation is already to adjust the responses of the organism. Greater flexibility of the programme thus allows certain types of aggression to be warded off. In the course of evolution, however, what has developed above all is the ability to collect outside information, to treat it and to adjust the reactions of the organism accordingly. All possible solutions are tried out, subject to the control of natural selection. Some organisms feel their environment, others hear it, or see it, or smell it. The ability to react to stimuli and the latitude in the choice of response increase in parallel. It is not enough to acquire a few impressions here or there; there must also exist the capacity to integrate them and to draw the conclusions. It is an advantage, for example, to be sensitive to light. So great is indeed this advantage that the eye has been 'invented' several times during evolution: the compound eye of insects, and the lens eye, which has arisen independently on at least three occasions: in certain molluscs, in arthropods and in the earliest vertebrates. But what would be the use of a precision instrument capable of defining shape, of judging distance, of determining the direction of movement, if not for locating a predator or a prey and making the appropriate response? For all that, it is essential to have the means of identifying the signals received, of comparing them with shapes recorded in a 'memory', of distinguishing friend from foe, of swimming, running or flying; in short, of 'choosing' a reaction. Means of perception, of reaction and of decision must evolve in harmony.

Increased exchanges between organism and environment are based on the development of the nervous system. But our present knowledge of this system is on a par with the knowledge of heredity in the nineteenth century. We have some information about certain electric or biochemical properties of the nerves; we have very little concerning specificity of the connections, or the organization and construction of the network. How is information coded, transmitted, recorded and deciphered? What logic underlies the activity of the brain, the memory or the acquisition of knowledge? In these areas, we are still almost completely ignorant. One fact seems beyond dispute, however: the anatomy of the nervous system is in some fashion fixed by heredity. The brain is like other organs: its structure is determined down to the last detail by the genetic programme. In many mutants of the mouse, alteration of one particular gene produces both an anomaly in behaviour and a specific lesion of the brain. During regeneration of severed nerves in certain organisms, the path adopted by the fibres, the establishment of the connections, the constitution of the circuits – in short, the entire organization of the network is carried out according to the original plan. Special centres exist, in fact, in the brain of mammals, not only for receiving various sensations and putting various muscles into action, but also for controlling sleep, or dreams, or attention, even for producing affective states. For example, there is a centre for 'punishment' in the rat, another for 'pleasure': fitted with correctly implanted electrodes and given the means of activating this centre at will, a rat satisfies itself until it collapses from sheer exhaustion! But we do not yet know how acquired circuits are superimposed on the heredity network, nor how the innate and the acquired fit together. For today the latter two are no longer antagonistic, but complementary. Ethologists consider that when behaviour involves acquired experience, it is dependent on the genetic programme. Learning comes into the framework fixed by heredity. Undoubtedly, it will soon become possible to analyse the molecular mechanism of synapses, the articulation of the nervous cells, the unity of anatomical connection on which is based the whole arrangement of the nervous system. And we can be sure that to the biochemist the characteristic reactions of brain activity will appear as ordinary as digestive reactions. But it is quite another matter to describe a feeling, a decision, a memory, a guilty conscience in terms of physics and chemistry. There is nothing to show that it will ever become possible, not only because of the complexity, but also because since Gödel we know that a logical system is not sufficient for its own description.

With the development of the nervous system, with learning and memory, the rigour of heredity is relaxed. In the genetic programme underlying the characteristics of a fairly complex organism, there is a closed part that can only be expressed in a fixed way and another open part that allows the individual a certain freedom of response. On the one hand, the programme rigidly prescribes structures, functions and attributes; on the other, it merely determines potentials, norms, frameworks. Here it commands; there it permits. As what is acquired increases in importance, the behaviour of the individual changes. This is apparent in the various ways by which birds recognize their like. In some, like the cuckoo, the species is identified in a way rigidly laid down in the genetic programme. All that is necessary is the mere sight of shapes and movements. Raised in the nest of its adoptive parents such as hedge-sparrows or warblers, as soon as the young cuckoo becomes independent it joins company with other cuckoos, even if it has never seen any in its life. With geese, on the contrary, identification is much more subtle. It works through a mechanism that ethologists call 'imprinting'. After hatching, the young gosling follows the first object it sees moving and hears calling. Usually, it is its real mother that the gosling follows. But if, by chance, it is another organism, Konrad Lorenz, for instance, then the gosling considers Konrad Lorenz as its mother and follows him everywhere. The genetic programme, therefore, determines shape in one case, and the ability to receive the imprint of a shape in the other. The animal world contains many examples of this kind. The growing importance of the open part of the programme gives a direction to evolution. Together with the capacity of response to stimuli, the degrees of freedom left to the organism in the choice of responses also increase. In man, the number of possible responses becomes so high that one can speak of the 'free will' so dear to philosophers. But flexibility has its limits. Even when the programme gives the organism only an ability, that of learning, for instance, it imposes restrictions on what can be learnt, on when learning is to take place and under what conditions. The genetic programme of man gives him an aptitude for language. It gives him the power of learning, understanding and speaking any language. But man must still be in a favourable environment at a certain stage of his development in order to fulfil this potential. After a certain age, deprived of speech, of care and of maternal affection for too long, a child will never learn to speak. The same restrictions apply to memory. There are limits to the amount of information that can be recorded, to the length of time it can be stored and to the power to retrieve it at will. But this boundary between rigidity and flexibility in the programme has hardly yet been explored.

As exchanges increase during evolution, new systems of communication appear that no longer operate within the organism, but between organisms. A network of relations is thus established between individuals belonging to the same species. Originally, these com-

THE LOGIC OF LIFE

munication systems were directly connected with the purpose of reproduction. Without them, sexuality would scarcely be efficacious. As long as it is not necessary for reproduction and remains merely an auxiliary function, nothing favours the union of the sexes. There is no 'sex-appeal' among bacteria. Opposite sexes meet by chance in random collisions. So it is with certain lower organisms that, being hermaphrodite, use sexual intercourse only occasionally. But as the independence of the organism increases, as sexuality becomes the only method of reproduction, then individuals of one sex must have a way of spotting those of the other. So long-distance communication systems appear that link selectively the opposite sexes of the same species. They are usually specific signals emitted by one sex and received by the other. Some insects use olfactory signals: they produce a volatile substance that is picked up, identified and interpreted by others provided by their genetic programme with a receptor sensitive to this molecular structure. Other insects use auditory signals: only the males sing. Fish and birds use visual signals: one of the sexes, usually the male, has a complex equipment of shapes, colours and iridescent ornaments that provide specific stimuli for the opposite sex. These visual signals, connected by hormones to the chemistry of the organism, activate all that part of behaviour concerned with reproduction. This sets off the succession of activities leading to copulation, nest-building, incubation and so on. There again, the whole series of operations, rites and ceremonies to be performed are all written down in the genetic message. The sight of the opposite sex acts merely as a signal. It only sets in motion the execution of an already prepared scheme for reproduction.

Obviously, these systems of signals have been selected to favour reproduction. They are nonetheless methods of communication between individuals of the same species. They make possible the formation of integrons at a higher level than the single organism. Up to mammals, however, integration rarely exceeds the temporary formation of a couple, the unit of reproduction. It is exceptional for groups of animals with coordinated behaviour to be set up, such as shoals of fish or flocks of birds during migration. The principal exception is found among certain insects, ants, termites or bees, that form true integrons transcending the individual. The old comparison of the organism and society becomes real in the ant-heap, the termitary and the beehive. Yet each of these structures is primarily a reproduction unit. The queen and the males play the role of sexual cells, the workers that of somatic cells. There again, the unity of these systems is rigidly determined by the genetic programmes that control, not only the morphology and physiology of each type, but also the nature and series of operations each has to perform. When the programme opens and a new system of communication such as the dance of the bees is established, it is in order to transmit information necessary for one function of the system: the search for food.

The structure of the genetic message therefore imposes the structure of animal communities. But with mammals the rigidity of the programme of heredity becomes less and less strict. The sense organs become refined. The means of action increase, particularly with the ability to grasp. The capacity to integrate becomes greater at the same time as the brain. One even sees the appearance of a new property: the ability to do without objects and interpose a kind of filter between the organism and its environment: the ability to symbolize. Gradually the signal becomes a sign. Even a rodent can learn to distinguish a triangle from a square or a circle and to associate shape with its quest for food. A cat can learn to count stimuli. Although a chimpanzee cannot speak with its larynx, it seems able to learn some elements of the code-language of gestures that deaf and dumb use for communication. The chimpanzee thus manages to recognize certain signs, interpreting and miming them, even combining some of them in groups to make short 'sentences' and express itself. It is not, therefore, at one stroke, by a sudden jump, that the small area of the brain controlling gesture and speech was developed. It is not even by a single series of successive stages, by a continuous chain, that man became man. It is by a mosaic of changes in which each organ, each system of organs, each group of functions, has evolved in its own way and at its own pace. Long foetal life and slow development, walking on two feet and freeing the forelimbs, formation of the hand and the use of tools, increase of brain size and aptitude for speech, all this leads not only to a greater autonomy with regard to environment, but also to new systems of communication, of regulation, of memory, which function at a higher level than the organism. All the conditions are then fulfilled for new integrations to occur in which coordination of elements no longer depends on the interaction of molecules, but on the exchange of coded messages. A new hierarchy of integrons is thus set up. From family organization to modern state, from ethnic group to coalition of nations, a whole series of integrations is based on a variety of cultural, moral, social, political, economic, military and religious codes. The history of mankind is more or less the history of these integrons and the way they form and change. There again appears a tendency towards growing integration made possible by the development of new means of communication. As long as it is confined to speech, the transfer of information is limited in space and time. With writing, communication can break free of time and the past experience of each individual can be stored in a collective memory. With electronics, with the means of preserving picture and sound and transmitting them to any point on the globe at a moment's notice, all restrictions in time and space have disappeared.

In the cultural and social integrons, new objects appear which function according to principles unknown at lower levels. The concepts of democracy, property and wages are as void of meaning for a cell or an organism as the concepts of reproduction and natural selection for an isolated molecule. This means that biology is diluted out in the study of man, just as is physics in the study of the cell. In this domain, biology represents merely one approach among others. Since the appearance of a theory of evolution, sociologists – starting with Herbert Spencer – have often tried to interpret the variations and interactions of social or cultural integrons by means of purely biological models. As the mechanisms governing the transfer of information obey certain principles, the transmission of culture through generations can be considered as a kind of second genetic system superimposed on heredity. It then becomes tempting, particularly for biologists, to compare the processes at work in both systems and draw analogies; to compare the ways ideas and mutations crop up; to contrast the novelty of change with the conservatism of the copy; to explain the disappearance of societies or cultures, like that of species, by blind alleys due to over-specialized evolution. The parallel can even be made in detail. Reproduction, then, lies at the centre of both systems, for codes of culture and societies as for the structure and properties of organisms: the fusion of cultures is like that of gametes; the university in society plays the role of the germ line in the species; ideas invade minds as viruses invade cells; they multiply and are selected for the advantages they confer on the group. In short, the variation of societies and cultures comes to be based on evolution, like that of species. All that has to be done, then, is to define the criteria of selection. The trouble is that no one has yet succeeded.

For with their codes, their regulations, their interactions, the objects that form cultural and social integrons transcend the explanatory schemes of biology. Once more, they involve integration of components that are themselves already integrated. But although there are further stages and discontinuities of phenomena and concepts, there is no complete break with the levels of biology. The objects of observation fit one inside the other. Physiology, for example, studies individually the functions of the organism and their coordinating mechanisms. At the level above physiology, behavioural science disregards the internal processes so as to grasp the complete reaction of the organism to its environment. At a still higher level, the dynamics of populations and sociology ignore the behaviour of individuals and study the behaviour of whole groups. One day, the different levels of observation will have to be brought together and related to each other. Once again, there is no hope of grasping the system without understanding the properties of its components. This means that although the study of man and societies cannot be reduced to biology alone, it cannot do without biology any more than biology can do without physics. It is not possible to account for cultural and social transformations by a selection of ideas. But it is not possible either to forget that the human organism is the product of natural selection. Of all living organisms, man has the most open and flexible genetic programme. But how flexible is it? How far is behaviour dictated by the genes? What constraints does heredity impose on the human mind? Obviously such constraints exist at some levels; but where should limits be drawn? According to modern linguistics, there is a basic grammar common to all languages; this uniformity would reflect a framework imposed by heredity on the organization of the brain. According to neurophysiologists, dreaming constitutes a necessary function not only for man, but for all mammals; it is controlled by a centre located in a precise area of the brain. According to ethologists, aggressiveness represents a form of behaviour selected in the course of evolution. Already present in most vertebrates, it gave man a selective advantage when living in small groups and constantly competing for food, women and power. Today, it is no longer natural selection that plays the leading role in transforming man, at least in certain societies. It is culture, more efficient, more rapid, but also very recent. Consequently, many aspects of man's behaviour today find their origin in some selective advantage given to the species when it emerged. Many traits of human nature must be inserted in the framework established by the twenty-three pairs of chromosomes that make up the common inheritance of man. But how rigid is this framework? What restrictions does the genetic programme impose on the plasticity of the human mind?

With the accumulation of knowledge, man has become the first product of evolution capable of controlling evolution. Not only the evolution of others, by encouraging species of interest to him and eliminating bothersome ones, but also his own evolution. Perhaps one day it will become possible to intervene in the execution of the genetic programme, or even in its structure, to correct some faults and slip in supplementary instructions. Perhaps it will also be possible to produce at will, and in as many copies as required, exact duplicates of individuals, a politician, for instance, an artist, a beauty queen or an athlete. There is nothing to prevent immediate application to human beings of the selection processes used for race-horses, laboratory mice or milch cows. But it seems desirable to know first the genetic factors involved in such complex qualities as originality, beauty or physical endurance. And above all, agreement has to be reached about the criteria for the choice. But that is no longer the concern of biology alone.

*

There is a coherence in the descriptions of science, a unity in its explanations, that reflects an underlying unity in the entities and principles involved. Whatever their level, the objects of analysis are always organizations, systems. Each of them is used as an ingredient by the one above. Even that old irreducible protagonist, the atom, has become a system. And physicists still cannot say whether the smallest entity known today is an organization or not. The word 'evolution' describes the changes that occur between systems. For what evolves is not matter blended with energy into one permanent whole. It is organization, the unit of emergence, that can always associate with its like to integrate into a system by which it is dominated. Without this property, the universe would be insipid: an ocean of identical particles, both inert and unaware of each other; something like the oldest rocks on earth, whose molecules and relationships have not changed for thousands of millions of years.

Integration changes the quality of things. For an organization often possesses properties that do not exist at the level below. These properties can be *explained* by the properties of the components; they cannot be *deduced* from them. This means that a particular integron has only a certain probability of appearing. All forecasts about its existence can only be statistical. This applies equally to the formation of beings and things; to the constitution of a cell, an organism or a population, as well as of a molecule, a stone or a storm. It is therefore on contingency that the unity of explanation is based today. In organisms, however, the effects of chance are immediately corrected by the requirements of adaptation, reproduction and natural selection. Hence a paradox. For in the inanimate world, the chances of events occurring can be statistically predicted with accuracy. In contrast, with living beings, which are indissolubly linked to a history whose details will never be known, the deviations introduced by natural selection make any prediction impossible. How could the appearance and development of certain living forms rather than others be predicted? How could, in the Secondary era, the sudden end of the large reptiles and the success of mammals have been foreseen?

Ultimately all organizations, all systems, all hierarchies owe their very possibility of existence to the properties of the atoms described by Clerk Maxwell's electromagnetic laws. There are perhaps other possible coherences in descriptions. But science is enclosed in its own explanatory system, and cannot escape from it. Today the world is messages, codes and information. Tomorrow what analysis will break down our objects to reconstitute them in a new space? What new Russian doll will emerge?

Notes

INTRODUCTION: THE PROGRAMME

1. C. Bernard, *Leçons sur les phénomènes de la vie*, Paris, 1878, vol. I, pp. 50–51.

CHAPTER ONE: THE VISIBLE STRUCTURE

1. Ambroise Paré, *Œuvres*, Paris, 1841, vol. III, p. 43.
2. Fernel, *De Abditis Rerum Causis*, I, 1; *Opera*, Geneva, 1637, pp. 483–4.
3. J. B. Porta, *Phytognomonica*, I, 8, Rouen, 1650, p. 14.
4. P. Belon, *La nature et diversité des poissons*, Paris, 1555, p. 87.
5. Paracelsus, *Les cinq livres de Auréole Philippe Théophraste de Hohenheim*, Prologue; French translation in *Œuvres médicales*, Paris, 1968, p. 194.
6. Paracelsus, *Le Livre Paragranum*, ibid., pp. 95–6.
7. Cesalpino, *De Plantis Libri XVI*, I, 14, Florence, 1583, p. 28.
8. Cardan, *Les Livres intitulés de la subtilité*, French translation, Paris, 1584, VIII, p. 196b.
9. Cesalpino, op. cit., I, p. 3.
10. Fernel, *Ambiani Physiologiae Libri*, IV, 1; *Opera*, p. 101.
11. Fernel, *De Abditis Rerum Causis*, II, 7; *Opera*, p. 590.
12. Montaigne, *Essays*, I, 8, tr. E. J. Trechmann, Oxford University Press, 1927, vol. I, p. 26.
13. Fernel, op. cit., I, 8, *Opera*, p. 538.
14. Cardan, *De la subtilité*, IX, p. 235b.
15. Paré, *De la génération de l'homme*, Preface; *Œuvres*, vol. II, p. 634.
16. Fernel, op. cit., I, 7; *Opera*, p. 535.
17. Paré, op. cit., X; *Œuvres*, vol. II, p. 651.
18. Fernel, op. cit., I, 3; *Opera*, p. 491.
19. Paré, op. cit., p. 638.
20. Montaigne, op. cit., II, 37, vol. II, p. 214.
21. Paracelsus, *Les cinq livres de Auréole Philippe Théophraste de Hohenheim*, III; *Œuvres médicales*, p. 222.
22. Fernel, *Ambiani Physiologiae Libri*, VII; *Opera*, p. 239.
23. Paré, *Des Monstres et Prodiges*, XVI; *Œuvres*, vol. III, p. 34.
24. Paré, op. cit., XX; *Œuvres*, vol. III, p. 44.
25. Paré, *Œuvres*, vol. III, Preface, p. 1.

26. Montaigne, op. cit., II, 30; vol. II, p. 161.
27. Descartes, *Rules for the Direction of the Mind*, XII; *The Philosophical Works of Descartes*, tr. E. S. Haldane & G. R. T. Ross, Cambridge University Press, 1931, p. 35.
28. Galileo, *Il Saggiatore, Opere*, Florence, 1890–1909, vol. VI, p. 232.
29. Descartes, op. cit., X; in *The Philosophical Works of Descartes*, p. 31.
30. Leibniz, *New Essays Concerning Human Understanding*, IV, XII, tr. Alfred Gideon Langley, Chicago and London, The Open Court Publishing Company, 1916, p. 526.
31. Descartes, *Principles of Philosophy*, IV, 205; *The Philosophical Works of Descartes*, p. 301.
32. Descartes, *Meditations*, I; *Discourse on Method and the Meditations*, tr. F. E. Sutcliffe, Penguin Books, 1968, p. 100.
33. Galileo, op. cit., p. 232.
34. Condillac, *Essai sur l'origine des connaissances humaines*, IV, 1, Amsterdam, 1746, vol. I, p. 179.
35. Descartes, *Principles of Philosophy*, IV, 203, loc. cit., p. 300.
36. Buffon, *De la manière d'étudier et de traiter l'histoire naturelle, Œuvres complètes*, Paris, 1774–9, [hereafter: (a)] vol. I, p. 17.
37. C. Bonnet, *Contemplation de la nature*, Amsterdam, 1766, p. 25; *Œuvres complètes*, Neuchâtel, 1781, vol. VII, p. 42.
38. ibid., p. 37; *Œuvres complètes*, vol. VII, pp. 79–81.
39. Galileo, *Discours concernant deux sciences nouvelles*, French translation, Paris, 1970, p. 107.
40. Borelli, *De Motu Animalium*, CCIV, Rome, 1685, p. 243.
41. Harvey, *On the Motion of the Heart and the Blood in Animals*, 1628, ch. XVII; *The Works of W. Harvey*, Sydenham Society, London, 1847 (reprinted 1965), pp. 78–80; Everyman edn, Dent, 1963, pp. 101–5.
42. Descartes, *Treatise on Man*; *Selections*, ed. Ralph M. Eaton, New York, Charles Scribner's Sons, 1927, p. 354.
43. ibid., p. 354.
44. Descartes, *Discourse on Method and the Meditations*, Penguin Books, 1968.
45. Réaumur, *Des Gâteaux de cire. Mémoires pour servir . . .; Morceaux choisis*, Paris, 1939, pp. 101–2.
46. Fontenelle, *Histoire de l'Academie Royale*, 1739, p. 35.
47. Buffon, *Discours sur la nature des animaux; Œuvres complètes* (a), vol. V, pp. 380–81.
48. Hartsoeker, *Cours de physique*, The Hague, 1730, vol. VII, p. 71.

49. Stahl, *Recherche sur la différence entre machine et organisme*, XXXIV; *Œuvres médico-philosophiques et pratiques*, Paris, 1859–63, vol. II, p. 289.

50. Hartsoeker, *Suite des éclaircissements sur les conjectures physiques*, Amsterdam, 1712, p. 55.

51. Geoffroy, *Table des différents rapports observés en Chymie: Traité de la Matière médicale*, Paris, 1743, vol. I, p. 18.

52. Lavoisier, *Traité de chimie*, Paris, 2nd edn, 1793, p. vi.

53. ibid., pp. xviii–xix.

54. ibid., p. xx.

55. ibid., p. xxi.

56. Réaumur, *Second mémoire sur la digestion*, Mém. Acad. Sc. Paris, 1752; *Morceaux choisis*, p. 202.

57. Lavoisier, *Premier mémoire sur la respiration des animaux* (Seguin et Lavoisier); *Œuvres*, Paris, 1862, vol. II, p. 691.

58. ibid., p. 697.

59. ibid., p. 700.

60. Tournefort, *Élémens de botanique*, Paris, 1694, vol. I, p. 4.

61. Linnaeus, *The Elements of Botany (Philosophia Botanica)*, London, 1775, p. 338.

62. Tournefort, *Introduction à la botanique*, in 'Tournefort', Museum National d'Histoire Naturelle, Paris, 1957, p. 284.

63. Linnaeus, op. cit., p. 335.

64. Tournefort, op. cit., vol. I, part 2, p. 47.

65. Linnaeus, op. cit., p. 364.

66. Buffon, *Les Pingouins et les Manchots*, *Œuvres complètes*, 1853–7, [hereafter: (b)] vol. VIII, p. 589.

67. Robinet, *De la Nature*, Amsterdam, 1766, vol. IV, p. 5.

68. Bonnet, *Contemplation de la Nature*, II, 10; *Œuvres*, vol. VII, p. 52.

69. ibid., *Œuvres*, vol. VII, p. 51.

70. Adanson, *Familles des plantes*, Paris, 1763, p. clxiv.

71. Buffon, *De la manière d'étudier...*; *Œuvres complètes* (a), vol. I, p. 54.

72. Tournefort, *Élémens de botanique*, p. 13.

73. Fontenelle, *Œuvres*, Paris, 1767, vol. V, p. 219–20.

74. Linnaeus, op. cit., p. 307.

75. Tournefort, *Introduction à la botanique*, p. 284.

76. Linnaeus, op. cit., pp. 230–31.

77. ibid., p. 232.

78. Tournefort, op. cit., p. 289.

79. Adanson, op. cit., p. clviii.

80. Buffon, *De la manière d'étudier...*; *Œuvres complètes* (a), vol. I, p. 88.

81. Tournefort, *Elémens de botanique*, p. 1.
82. Linnaeus, op. cit., p. 264.
83. ibid., p. 304.
84. Tournefort, *Introduction à la botanique*, p. 297.
85. Linnaeus, op. cit., p. 305.
86. John Ray, *Historia Plantarum*, London, 1686, vol. I, ch. xx, pp. 40–41.
87. Buffon, *Histoire naturelle des animaux*; *Œuvres complètes* (a), vol. III, p. 15.
88. Linnaeus, op. cit., p. 232.
89. Buffon, *De la Nature*; *Œuvres complètes* (b), vol. III, p. 414.
90. Harvey, *On the Generation of Animals*, 1651; *The Works of W. Harvey*, Sydenham Society, London, 1847 (reprinted 1965), *On Parturition*, p. 575.
91. ibid., p. 577.
92. Descartes, *Primae Cogitationes*; *Œuvres*, Paris, 1897–1913, vol. XI, p. 506.
93. Descartes, *Formation de l'animal*; *Œuvres*, vol. XI, p. 277.
94. Malebranche, *Entretiens sur la metaphysique, sur la religion et sur la mort*. Paris, 1711, vol. II, pp. 14–15.
95. Redi, *Experimenta circa Generationem Insectorum*; *Opera*, Amsterdam, 1686, vol. I, pp. 30–31.
96. ibid., pp. 17–18.
97. Steno, *Elementorum Myologiae Specimen*, Florence, 1667, p. 117.
98. Regnier de Graaf, *Traité des parties des femmes qui servent à la génération*, Warsaw, 1701, Preface, pp. ii–iii.
99. Leeuwenhoek, Letter to Grew, 25 April 1679; *Arcana naturae*, Leyden, 1696, part II, p. ii.
100. Malebranche, *Entretiens* . . . vol. II, p. 13.
101. Hartsoeker, *Essay de dioptrique*, ch. X, 89, Paris, 1694, p. 229.
102. ibid., p. 228.
103. Leeuwenhoek, Letter to Leïbniz; *Epistoles physiologicae*, Delft, 1719, p. 294.
104. Hartsoeker, op. cit., p. 230.
105. Geoffroy, Thèse de médecine: *Traité de la Matière médicale*, vol. I, p. 95.
106. Malebranche, *Recherche de la Vérité*, Paris, 1700, vol. I, p. 48.
107. Perrault, *De la mécanique des Animaux*; *Œuvres diverses de physique et de mécanique*, Leyden, 1721, p. 485.
108. Malebranche, *Recherche de la Vérité*, vol. I, pp. 47–8.
109. Swammerdam, *Histoire générale des Insectes*, Utrecht, 1682, p. 48.

110. Malebranche, op. cit., vol. I, p. 48.
111. Hartsoeker, op. cit., p. 231.
112. Malebranche, op. cit., vol. I, p. 200.
113. Leibniz, *Theodicy*, Routledge & Kegan Paul, 1951, p. 172.
114. Fontenelle, *Lettres galantes*; *Œuvres*, vol. I, p. 322–3.
115. Haller, cited in Bonnet, *Considération sur les corps organisés*; *Œuvres*, vol. VI, p. 443.
116. Bonnet, *Contemplation de la Nature*, VIII, 8; *Œuvres*, Amsterdam, 1766, vol. VIII, p. 130.
117. Bonnet, *Insectologie*, VIII, 10; *Œuvres*, vol. I, p. 230.
118. Bonnet, *Contemplation de la Nature*; *Œuvres*, vol. VIII, p. 71.
119. Réaumur, *Mémoire sur les grenouilles*; *Morceaux choisis*, p. 250.
120. ibid., p. 247.
121. Spallanzani, *Expériences pour servir à l'histoire de la génération*, Geneva, 1785, XIII, p. 13.
122. ibid., CXIX, pp. 128–9.
123. ibid., CLXIX, p. 201.
124. ibid., CLII, p. 184.
125. Buffon, *Histoire naturelle des Animaux*; *Œuvres complètes* (a), vol. III, pp. 231–2.
126. Réaumur, *Mémoires Acad. Sc.*, Paris, 1712, p. 235.
127. Maupertuis, *Vénus Physique*, *Œuvres*, Lyons, 1768, vol. II, p. 70.
128. Bonnet, *Lettres*, *Œuvres*, vol. XII, p. 382.
129. Réaumur, *Art de faire éclore les Poulets*, 2nd edn, Paris, 1751, vol. II, p. 340.
130. ibid., pp. 366–7.
131. Koelreuter, cited in R. C. Olby, *Origins of Mendelism*, Constable, 1966, p. 154.
132. Maupertuis, *Lettres*; *Œuvres*, vol. II, pp. 307–8.
133. Réaumur, op. cit., pp. 377 ff.
134. Bonnet, *Corps organisés*, CCCLV; *Œuvres*, vol. VI, p. 479.
135. Maupertuis, op. cit., pp. 309–10.
136. Newton, *Opticks*, G. Bell & Sons, 1931, p. 1.
137. Réaumur, *Travaux de l'Academie des Sciences*, 1712, p. 226.
138. Buffon, *Histoire naturelle des Animaux*; *Œuvres complètes* (a), vol. III, p. 25.
139. Bonnet, *Œuvres*, vol. XV, p. 356.
140. Maupertuis, *Lettres*; *Œuvres*, vol. II, p. 418.
141. Buffon, *Histoire naturelle des Animaux*; *Œuvres complètes* (a), vol. III, p. 48.

Notes

CHAPTER TWO: ORGANIZATION

1. Haller, *Éléments de physiologie*, pt I, ch. 1, x, French translation, Amsterdam, 1769, p. 3.
2. ibid., p. 2.
3. Bonnet, *Palingénésie philosophique*, IX, 1; *Œuvres*, vol. XV, p. 350.
4. Buffon, *Histoire naturelle des Animaux*; *Œuvres complètes* (a), vol. III, pp. 27–8.
5. Maupertuis, *Vénus physique*; *Œuvres*, vol. II, p. 120.
6. Buffon, op. cit., pp. 34–5.
7. Maupertuis, op. cit., p. 121.
8. Bonnet, *Corps organisés*; *Œuvres*, vol. VI, pp. 390–91.
9. Maupertuis, op. cit., pp. 121–2.
10. Maupertuis, *Système de la Nature*, XIX; *Œuvres*, vol. II, p. 149.
11. Maupertuis, op. cit., XXXIII; *Œuvres*, vol. II, pp. 158–9.
12. Buffon, op. cit., pp. 48–9.
13. ibid., p. 51.
14. ibid., p. 51.
15. ibid., p. 61.
16. Buffon, *Le Cerf*; *Œuvres complètes* (b), vol. II, p. 521.
17. Maupertuis, op. cit., p. 132.
18. Buffon, *Œuvres complètes* (a), vol. III, p. 450.
19. Buffon, *Introduction à l'histoire des minéraux*; *Œuvres complètes* (a), vol. VI, p. 24.
20. Buffon, *Nomenclature des Singes*; *Œuvres complètes* (b), vol. IV, p. 15.
21. Lamarck, *Philosophie zoologique*, Paris, 1873, vol. I, p. 62.
22. Goethe, *Introduction générale à l'anatomie comparée*; *Œuvres d'histoire naturelle*, French translation, Paris, undated, p. 26.
23. de Jussieu, *Principes de la méthode naturelle des végétaux*, from the *Dictionnaire des sciences naturelles*, Paris 1824, p. 27.
24. Lamarck, *La Flore française*, Paris 1778, vol. I, p. xcvii.
25. Lamarck, *Philosophie zoologique*, vol. I, p. 102.
26. ibid., p. 63.
27. ibid., p. 119.
28. Lamarck, *Histoire naturelle des animaux sans vertèbres*, 1815–22, vol. I, pp. 130–31.
29. Lamarck, *Philosophie zoologique*, vol. I, p. 189.
30. Lamarck, *La Flore française*, vol. I, pp. 1-2.
31. Lamarck, *Philosophie zoologique*, vol. I, p. 106.
32. Lamarck, *Histoire naturelle des animaux sans vertèbres*, vol. I, pp. 49–50.

33. Kant, *The Critique of Judgement*, tr. James Creed Meredith, Oxford University Press, 1952, part II, pp. 17–18.
34. Goethe, op. cit., p. 30.
35. Kant, op. cit., p. 22.
36. Bichat, *Recherches physiologiques sur la vie et la mort*, 1810, p. 97.
37. Cuvier, *Leçons d'anatomie comparée*, 2nd edn, Paris, 1835, vol. I, p. 2.
38. Bichat, op. cit., p. 1.
39. Cuvier, op. cit., vol. I, p. 4.
40. Goethe, op. cit., p. 19.
41. Liebig, *Chimie organique appliquée à la physiologie animale*, French translation, Paris, 1842, p. 209.
42. Cuvier, loc. cit.
43. Bichat, *Anatomie générale*, 2nd edn, Paris, 1818, vol. I, p. 17.
44. Cabanis, *Rapport du physique et du moral*, Paris, 1830, vol. II, p. 256.
45. Liebig, op. cit., Preface, p. ix.
46. ibid., p. 171.
47. Wöhler, cited in J. Loeb, *The Dynamics of Living Matter*, Columbia University Press, 1906, New York, p. 7.
48. Liebig, op. cit., pp. 202–3.
49. ibid., p. 215.
50. ibid., p. 218.
51. ibid., p. 112.
52. Liebig, *Chimie appliquée à la physiologie végétale*, French translation, Paris, 1844, p. 365.
53. Berzelius, cited in J. Loeb, *The Dynamics of Living Matter*, pp. 7–8.
54. Cuvier, Letter to Mortrud; *Leçons d'anatomie comparée*, vol. I, p. xvii.
55. Cuvier, *La Règne animal distribué d'après son organisation*, 1817, vol. I, p. 7.
56. Goethe, op. cit., p. 19.
57. E. Geoffroy Saint-Hilaire, *Philosophie anatomique*, Paris, 1818, pp. xxii–xxiii.
58. Cuvier, *Leçons d'anatomie comparée*, 1, p. 50.
59. E. Geoffroy Saint-Hilaire, op. cit., p. xxx.
60. ibid., p. 19.
61. Goethe, op. cit., p. 30.
62. E. Geoffroy Saint-Hilaire, *Considérations sur les pièces de la tête osseuse*, in *Ann. Mus. Nat.*, 1807, 10, p. 342.
63. Cuvier, *Le Règne animal*, vol. I, Preface, p. iv.
64. Cuvier, *Leçons d'anatomie comparée*, vol. I, p. 49.

65. ibid., vol. I, pp. 56–7.

66. ibid., vol. I, p. 58.

67. Cuvier, *Le Règne animal*, vol. I, p. 10.

68. ibid., pp. 55–6.

69. E. Geoffroy Saint-Hilaire, *Philosophie anatomique*, pp. 18–19.

70. E. Geoffroy Saint-Hilaire, *Mémoire sur l'organisation des insectes*, in *Journal compl. des sciences médicales*, 1819, 5, p. 347.

71. Cuvier, op. cit., vol. I, p. 6.

72. Cuvier, *Leçons d'anatomie comparée*, vol. I, p. 60.

73. Cuvier, *Le Règne animal*, vol. I, p. 57.

74. Cuvier, *Mémoire sur les céphalopodes*, Paris, 1817, 1. 143.

75. Cuvier, *Leçons d'anatomie comparée*, vol. I, p. 59.

76. Bichat, *Traité des membranes*, Paris, 1816, p. 29.

77. ibid., p. 28.

78. ibid., p. 31.

79. Bichat, *Anatomie générale*, p. 35.

80. Oken, cited in M. Klein, *Histoire des origines de la théorie cellulaire*, Paris, 1936, p. 19.

81. Dujardin, *Recherches sur les organismes inférieurs*, in *Ann. sc. naturelles*, 1835, 2nd series, vol. IV, p. 367.

82. Schwann, *Microscopical researches into the accordance in the structure and growth of animals and plants*, tr. H. Smith in *Schwann and Schleiden Researches*, Sydenham Society, 1847, p. 165. [Reprinted in *Great Experiments in Biology*, ed. M. L. Gabriel and S. Fogel, Prentice-Hall, New York, 1955.]

83. ibid., p. 192.

84. Schleiden, *Contributions to Phytogenesis*, tr. H. Smith in *Schwann and Schleiden Researches*, 1847, pp. 231–2.

85. Schwann, loc cit., p. 193.

86. ibid., p. 199.

87. ibid., p. 190.

88. Virchow, *Die Cellularpathologie in ihrer Begründung auf physiologische und pathologische Gewebelehre*, Berlin, 1858, p. 25.

89. ibid., p. 12.

90. von Baer, cited in E. Haeckel, *Anthropogénie*, French translation, Paris, 1877, p. 165.

91. von Baer, cited in Darwin, *The Origin of Species*, 6th edn, 1872, reprinted in World's Classics, Oxford University Press, 1951, p. 506.

92. E. Geoffroy Saint-Hilaire, *Philosophie anatomique*, II, *Des monstruosités humaines*, Paris, 1822, p. 539.

93. I. Geoffroy Saint-Hilaire, *Histoire Générale et Particulière des*

anomalies de l'organisation chez l'homme et les animaux, Paris, 1832, vol. I, 1, 18.

94. Remak, *Untersuchungen über die Entwicklung der Wirbelthiere*, Berlin, 1850, p. 140.

95. von Baer, cited in E. B. Wilson, *The Cell in Development and Heredity*, Macmillan, New York, 1925, p. 1035.

96. Virchow, op. cit., p. 25.

97. Comte, *Cours de philosophie positive*; *Œuvres*, Paris, 1838, vol. III, p. 237.

CHAPTER THREE: TIME

1. Buffon, *Théorie de la terre*; *Œuvres complètes*, vol. I, pp. 109–11.
2. ibid., vol. II, p. 2.
3. Diderot, *Entretien entre d'Alembert et Diderot*, edn de la Pleiade, 1946, p. 907; tr. J. Stewart and J. Kemp, in *Diderot: Selected Writings*, International Publishers, New York (Lawrence & Wishart, London), 1963, pp. 53–4.
4. de Maillet, *Telliamed*, The Hague, 1755, vol. II, p. 166.
5. ibid., p. 169.
6. ibid., p. 171.
7. Robinet, *De la Nature*, vol. IV, p. 17.
8. Bonnet, *Palingénésie philosophique*; *Œuvres*, vol. XV, p. 192.
9. ibid., pp. 219–20.
10. Buffon, *Du Lièvre*; *Œuvres complètes* (b), vol. II, p. 540.
11. Buffon, *Supplément à la théorie de la terre*; *Œuvres complètes* (b), vol. IX, p. 424.
12. Buffon, *L'Unau et l'Aï*; *Œuvres complètes* (b), vol. III, p. 443.
13. Diderot, *Lettre sur les aveugles*, edn de la Pléiade, 1946, p. 871.
14. Buffon, *Vue de la nature*, 2e vue; *Œuvres complètes* (b), vol. III, p. 414.
15. Buffon, *Dégénération des animaux*; *Œuvres complètes* (b), vol. IV, p. 123.
16. Buffon, *Vue de la nature*, 2e vue; *Œuvres complètes* (b), vol. III, p. 418.
17. Buffon, *Dégénération des animaux*; *Œuvres complètes* (b), vol. IV, p. 144.
18. Maupertuis, *Système de la nature*; *Œuvres*, vol. III, p. 164.
19. Maupertuis, *Vénus physique*; *Œuvres*, vol. II, p. 110.
20. ibid., pp. 110–11.
21. Maupertuis, *Essai de cosmologie*; *Œuvres*, vol. I, p. 11.
22. ibid.
23. Lamarck, *Philosophie zoologique*, vol. I, p. 122.
24. ibid., p. 124.

25. ibid., pp. 2–3.
26. ibid., p. 83.
27. ibid., pp. 81–2.
28. ibid., p. 231.
29. ibid., p. 103.
30. ibid., p. 268.
31. ibid., p. 29.
32. ibid., p. 168.
33. ibid., p. 144.
34. ibid., p. 28.
35. ibid., p. 226.
36. ibid., pp. 222–3.
37. Lamarck, op. cit., vol. I, p. 214, vol. II, p. 81.
38. Lamarck, op. cit., vol. I, p. 214, vol. II, pp. 81–2.
39. Cuvier, *Leçons d'anatomie comparée*, vol. I, p. 59.
40. Comte, *Cours de philosophie positive*; *Œuvres*, vol. III, p. 235.
41. Cuvier, *Discours sur les révolutions de la surface du globe*, 1830, p. 14.
42. ibid., p. 18.
43. Lyell, *Principles of Geology*, 6th edn, John Murray, 1840, vol. I, Preface, p. ix.
44. Lyell, op. cit., vol. I, p. 325.
45. Lyell, *Manual of Elementary Geology*, 3rd edn, John Murray, 1851, p. 98.
46. ibid., p. 94.
47. ibid., p. 94.
48. ibid., pp. 109–10.
49. ibid., p. 501.
50. Humboldt, *Cosmos*, French translation, Paris, 1855, pp. 316–17.
51. Darwin, *The Origin of Species*, [1st edn, 1859, John Murray; reprinted] Penguin Books, 1968, p. 384.
52. ibid., p. 444.
53. ibid., p. 172.
54. ibid., p. 413.
55. ibid., p. 404.
56. ibid., p. 395.
57. Wallace, *On the tendency of varieties to depart indefinitely from the original type*, in *Journal of the Linnaean Society*, 1859, vol. 3, p. 45. [Also in *Evolution by Natural Selection*, ed. G. de Beer, Cambridge University Press, 1958, pp. 268–79.]
58. Malthus, *Essay on the Principle of Population*, 7th edn, Everyman Library, Dent, 1967, pp. 5–6.

59. Darwin, op. cit., p. 98.
60. ibid., p. 131.
61. Darwin, *The Origin of Species* [6th edn, 1872, reprinted World's Classics], Oxford University Press, 1951, p. 101.
62. Darwin, *The Origin of Species* [1st edn, 1859, reprinted], Penguin Books, 1968, p. 162.
63. Darwin, *The Origin of Species* [6th edn, 1872, reprinted World's Classics], Oxford University Press, 1951, p. 133.
64. Darwin, *The Origin of Species* [1st edn, 1859, reprinted], Penguin Books, 1968, pp. 137–8.
65. Wallace, loc. cit. and *Evolution in Natural Selection*, p. 273.
66. Darwin, *The Origin of Species* [6th edn, 1872, reprinted World's Classics], Oxford University Press, 1951, p. 143.

CHAPTER FOUR: THE GENE

1. Bernard, *Leçons sur les phénomènes de la vie*, 1879, vol. II, p. 2.
2. Bernard, *Leçons de pathologie expérimentale*, 1872, p. 493.
3. Bernard, *Leçons sur les phénomènes de la vie*, vol. I, p. 356.
4. Bernard, *La Science expérimentale*, 1878, p. 70.
5. Bernard, *Introduction à l'étude de la médecine expérimentale*, 1865, p. 34.
6. Bernard, *Leçons sur les propriétés des tissus vivants*, 1866, p. 6.
7. Bernard, *Leçons sur les phénomènes de la vie*, vol. II, p. 5.
8. Bernard, op. cit., vol. I, p. 121.
9. ibid., vol. I, p. 335.
10. ibid., p. 342.
11. ibid., p. 332.
12. Bernard, *La Science expérimentale*, p. 133.
13. Bernard, *Leçons sur les phénomènes de la vie*, vol. I, p. 340.
14. Clerk Maxwell, *Theory of Heat*, 10th edn, London, 1891, p. 338.
15. Naudin, *Nouvelles recherches sur l'hybridité dans les végétaux*, in *Ann. Sc. Nat. Botanique*, 4th series, 19, p. 194.
16. Mendel, *Experiments in Plant Hybridization*, English translation in W. Bateson, *Mendel's Principles of Heredity*, Cambridge University Press, 1908, p. 318. [Reprinted in *Classic Papers in Genetics*, ed. J. A. Peters, Prentice-Hall, New York, 1959, p. 2.]
17. ibid., Bateson, p. 321, *Classic Papers*, p. 4.
18. ibid., Bateson, pp. 336–7, *Classic Papers*, pp. 13–14.
19. ibid., Bateson, p. 331, *Classic Papers*, p. 10.
20. ibid., Bateson, p. 344, *Classic Papers*, p. 18.
21. ibid., Bateson, p. 345, *Classic Papers*, p. 19.

22. Mendel, Letter to Carl Nägeli, 18 April 1867, reprinted in *Great Experiments in Biology*, p. 229.
23. Haeckel, *Anthropogénie*, French translation, Paris, 1877, p. 84.
24. Richard von Hertwig, *Die Protozoen und die Zelletheorie*, in *Archiv für Protistenkunde*, 1902, vol. I, p. 1.
25. Boveri, cited in F. Baltzer, *Theodor Boveri*, Berkeley, 1967, p. 121.
26. ibid., p. 70.
27. ibid., p. 68.
28. Boveri, cited in E. B. Wilson, *The Cell in Development and Heredity*, 1925, p. 916.
29. Huxley, *Evolution in Biology* (*Encyclopedia Britannica*, 9th edn, 1878), lecture XI in *Science and Culture*, London, 1882, p. 296.
30. Weismann, *Essais sur l'hérédité*, French translation, Paris, 1892, p. 171.
31. ibid., p. 124.
32. ibid., p. 125.
33. ibid., p. 318.
34. ibid., p. 154.
35. ibid., p. 176.
36. ibid., pp. 175–6.
37. de Vries, *Espèces et variétés*, French translation, Paris, 1909, p. 458.
38. Association of American Stockbreeders, *Proceedings*, vols. 1–5, 29 December 1903.
39. de Vries, *Espèces et variétés*, pp. 179–80.
40. ibid., p. 18.
41. ibid., p. 299.
42. ibid., p. 361.
43. Helmholtz, *On the application of the law of the conservation of force to organic Nature*, in *Notice of the Proceedings of the Royal Institution of Great Britain*, 12 April 1861, p. 357.
44. Berthelot, *La Synthèse chimique*, 1897, p. 203.
45. ibid., p. 215.
46. ibid., p. 240.
47. ibid., p. 272.
48. ibid., p. 277.
49. Pasteur, *Dissymétrie moléculaire*; *Œuvres*, 1922, vol. I, p. 327.
50. ibid., p. 364.
51. Pasteur, *Fermentations et générations dites spontanées*; *Œuvres*, vol. II, p. 224.
52. Pasteur, *Études sur la bière*; *Œuvres*, vol. V, p. 216.

53. Pasteur, *Fermentations et générations dites spontanées*; *Œuvres*, Vol. II, p. 224.

54. Büchner, *Berichte deutsche chemische Gesellschaft*, 1897, vol. 30, pp. 117–24, English translation in *Great Experiments in Biology*, p. 28.

55. Ostwald, *Lehrbuch der allgemeinen Chemie*, 1902, vol. II, p. 248.

56. Loeb, *The Dynamics of Living Matter*, Columbia University Press, New York, 1906, p. 29.

57. ibid., p. 33.

58. ibid., p. 37.

59. Haldane, *The Philosophical Basis of Biology*, London, 1931, p. 12.

60. Bohr, 'Light and Life' in *Nature*, London, 1933, vol. 131, p. 458.

61. Schrödinger, *What is Life?*, Cambridge University Press, 1956, p. 67.

CHAPTER FIVE: THE MOLECULE

1. Schrödinger, *What is Life?*, Cambridge University Press, 1956, p. 68.

2. Wiener, *The Human Use of Human Beings*, 1954, Doubleday, New York, p. 95.

3. ibid., p. 26.

4. Schrödinger, *What is Life?*, pp. 18–19.

5. J. Clerk Maxwell, in Wiener, *Cybernetics*, 2nd edn, 1961, Massachusetts Institute of Technology, p. 58.

Index

Figures in *bold type* indicate whole chapters or sections; '*p*' means *passim* (here and there) – scattered references.

Index

'Generation' (article in *Grande Encyclopédie*), 73; *quoted*, 73
genes (*see also* chromosomes, enzymes *and* genetics), 14, 16, 17, **178–246**, 261–5 *p*, 275, 284; experimentation, **180–92**; statistical analysis, **192–291**; birth of genetics, **201–9**; chromosomes, **209–26**; enzymes, **226–43**; and proteins, 243; synthetic, 295
genetic: activity, 264; analysis, 260–67 *p*; code, 3, 276–7, 305, 306 *p*; heritage, 1; information, 250–51, 291; material, 223–5 *p*, 264, 265, 290, 292; message (text), **267–78**, 287, 290–92 *p*, 297, 298 *p*, 319 *p*
genetic programme(s) (*see also* programme), 3 *p*, 8, 9, 287, 291–2 *p*, 298, 301, 304, 308–11 *p*, 314, 315–17 *p*, 319; origin of (?), 305; expansion of, 311–12; man's, 321; text of, *see* genetic message
genetics (and geneticists, *see also* genes), 180 *p*, 219, 220, 223–4 *p*, 226, 243–4 *p*, 249, 260–62 *p*, 288; birth of, **201–9**; classical, 225; aim and attitude of, 226; and chemistry, 226 (*see also* enzymes); common to all organisms, 263
genus, 50, 51
Geoffroy, 41, 59
geology, 132, 156–60
Gerhard, 99
germ cells, 215–17 *p*
'germ' (disease), 234
germs of organic bodies, 57–62, 71–3 *p*; 'so few able to develop', 167
Gibbs, 167, 173, 196, 197, 199–201 *p*
God: as Creator, 20, 22–31 *p*, 36, 62
Gödel, 316
Goethe, Johann Wolfgang von, 11, 86, 108, 152; *quoted*, 84, 89, 90, 101, 103
gravitation, 31, 40, 75
Grew, 112
'growth factor', 241 *p*

Haeckel, 191 *p*, 209; *quoted* 210

Haldane, J. S., *quoted*, 244
Haller, 64, 73, 75, 112, 150; *quoted*, 63, 75
Hartsoeker, 15, 39, 40; *quoted*, 39, 40, 59 *p*
Harvey, William, 14, 34, 42; *quoted* 34; and mechanism, 34–5; and generation, 35, 52–3
heat, 24–5; importance of, in organic bodies, 95–6
Helmholtz, 194; *quoted*, 228
hereditary tendency, 218
heredity (*see also* genetics *and* inheritance), 1–4 *p*, 8, 14, 25, 26, **67–73**, 139, 164, 176, 178, 180, 190–92 *p*, 201–2, 209, 215, 219–20 *p*, 222–3 *p*, 226, 244 *p*, 254, 262–3 *p*, 265, 295, 299, 315; influence of environment on, 2, 149; history of, 10; bilateral, 73; memory of, **75–82**, 180, 207, 254; 'creates but does not adapt', 149; perpetuates variation, 170, 172; Bernard on, 190; complexity of, 202; Mendel and a science of, 202–9; and cellular continuity, 210; and the individual, 216–18; and mathematics, 218; quantum theory of, 224; and evolution, 225; and computer memory, 254; of micro-organisms, 261; mechanics of, 261; and computers, 265; and the nervous system, learning and memory, 316
Hertwig: brothers, 213; O., 212; R., 211–12
hidden architecture, **82–8**
'hierarchy' of five levels (in C 18), 48–9
Hippocrates, 20, 206
histology (and histologists), 154, 180
Holbach, Paul H. D., baron de, 39
homology (and homologies), 101–2, 103, 104
honeycombs, 37–8
Hooke, Robert, 15, 112
hormones, 8, 185, 190, 247, 301, 312, 318
horses, 138; Androvandus on, 22; and flies, *compared*, 54

ABOUT THE AUTHOR

François Jacob was born in 1920 in Nancy, France. After attending the Lycée Carnot in Paris, he began studying medicine at the Faculty of Paris with the intention of becoming a surgeon. In June 1940 his studies were interrupted by the war; he left France to join the Free French Forces in London, was sent to Africa as a medical officer, was posted to the Second Armoured Division, and was severely wounded in Normandy in 1944.

After the war, François Jacob finished his medical studies, completing his doctorate in 1947. Unable to practice surgery because of his war injuries, he worked in various fields before turning to biology. He obtained his doctorate in science in 1954 at the Sorbonne. In 1956 he was appointed Laboratory Director of the Institut Pasteur, then in 1960 head of the Department of Cell Genetics. In 1964 he was also appointed Professor of Cell Genetics at the Collège de France. He is a foreign member of the Danish Royal Academy of Arts and Sciences, the American Academy of Arts and Sciences, the American Philosophical Society, the Royal Society, the Académie Royal de Médecine de Belgique, and the National Academy of Sciences of the United States.

François Jacob has worked mainly on the genetic mechanisms of bacteria and bacterial viruses, on the information transfer and regulatory system in the bacterial cell. In 1965, he was awarded jointly with André Lwoff and Jacques Monod the Nobel Prize for Medicine.